"*Southern Forages* is simply an all inclusive guide to forage selection, production, and utilization by livestock. It is a reference book that every forage producer should have."

Mr. Eugene Glenn, Beef Cattleman/Commercial Hay Producer, Decatur, Alabama

* * * * *

"I have found *Southern Forages* to be an excellent resource for students, farmers, and farm advisors. It is user friendly and its size, shape, and page quality make it appealing to a diverse audience."

Dr. Jim Green, Extension Forage Specialist, Nort

* * * * *

"Page after page reveals keystone principles of a dashboards of trucks to libraries, this book will b highly recommend *Southern Forages*."

Dr. Jimmy Henning, Extension Forage Specialist, University of Kentucky

* * * * *

"*Southern Forages* is a 'must read' for every livestock producer who uses pasture and hay or silage. I find it a rare blend of scientific and easy-to-read practical information. It is one of my most-used references."

Earl Manning, Mid-South Editor, *Progressive Farmer* magazine.

* * * * *

"Any livestock producer who is planning a new forage program should read this book. Any livestock producer with an established forage program who reads this book will be stimulated to reevaluate and improve that program. (It) is very well-organized, interesting, and easy to read."

Dr. John Moore, Animal Scientist/Nutritionist, University of Florida

* * * * *

"I like the approach used in separating the various types of forages as well as the various types of livestock. Too often, information concerning forages and pastures for horses, sheep, and wildlife is omitted from traditional texts. I found the book easy to read and suitable as either a textbook for the classroom or a handbook for the producer."

Dr. Monte Rouquette, Jr., Forage Physiologist, Texas A&M University Agricultural Research and Extension Center, Overton

* * * * *

"There are three books that I couldn't do without: the *Bible, Progressive Fallow Deer Farming,* and *Southern Forages.*"

Cleve Tedford, Deer Farmer, Tellico Plains, Tennessee

* * * * *

"At long last, a great book on how, when, what, and why to grow and use forages in the South has been written! A masterful job of blending complex scientific facts in everyday language."

Mr. Warren Thompson, National Forage Specialist, ABI Alfalfa, Lexington, Kentucky

* * * * *

"*Southern Forages* contains a unique combination of features not found in other texts. Examples are the full-color photographs which clearly illustrate characteristics of forage species, and the appendix which serves as a valuable compendium, summarizing key forage-related information."

Dr. J.J. Volenec, Forage Researcher/Teacher, Purdue University

Dedication . . .

To Vonda, Pat, and Cheryl

SOUTHERN FORAGES

SECOND EDITION

By Donald M. Ball, Carl S. Hoveland, and Garry D. Lacefield

Published by the Potash & Phosphate Institute (PPI)
and the Foundation for Agronomic Research (FAR)
655 Engineering Drive, Suite 110
Norcross, Georgia 30092-2843 USA

SOUTHERN FORAGES

SECOND EDITION
1996

Prior edition: 1991.

Typesetting, composition, and printing by
Williams Printing Company, Atlanta, Georgia, USA

Order from:
Potash & Phosphate Institute (PPI)
655 Engineering Drive, Suite 110
Norcross, Georgia 30092-2843 USA

Library of Congress Catalog Card No. 95-71361

ISBN 0-9629598-2-0

Foreword

When *Southern Forages* was first published in 1991, we had high expectations for its success. There were several reasons why, the most important being that the authors were all world-recognized forage scientists and offered us a great first manuscript. Further, the development phase had gone exceptionally well, publishers and authors in step from start to finish. Finally, the editorial group who put the book together had a proven track record.

Now, four years and thousands of book sales later, we are happy to report that our expectations have been met and surpassed. Our early assessment as to the quality and value of *Southern Forages* has been more than justified by the book's readers and users, from field to classroom. A growing number of universities and colleges have adopted the book as a text . . . acceptance has been virtually unanimous.

Now we are presenting the first updating of *Southern Forages*. While changes have been made to improve readability and utility, the fact remains that this edition carries an even greater wealth of information . . . in an easy-to-read, understandable format.

Some readers have suggested that the title of *Southern Forages* is too restrictive, because the book contains principles and concepts that have much broader adaptability. In fact, it is being utilized as a forage reference in several regions of the USA, as well as in other countries. For example, an effort is currently in progress to translate and adapt the book for use by forage workers in Poland.

We continue to be grateful to Drs. Ball, Hoveland and Lacefield for giving us the opportunity to partner with them on *Southern Forages*. We sincerely hope that you, the reader, will appreciate the book as much as we do.

D.W. Dibb, President
Potash & Phosphate Institute

B.C. Darst, President
Foundation for Agronomic Research

Preface

In the Preface to the First Edition of *Southern Forages*, we emphasized the great economic and environmental value of forages in the South. We mentioned the unrealized potential of forages, the opportunity for improvement which exists on most farms, and the need for a modern, practical book that contains basic information for successful forage production in the South.

Our objective in writing *Southern Forages* was to provide a book on practical aspects of forage production that any forage producer or agricultural worker, regardless of his or her scientific training, could read and understand. Our editor and publisher encouraged us to use color photographs and diagrams to illustrate plant species and concepts. The result was an attractive and easy to read book.

The response to our first edition of *Southern Forages* was gratifying. It was widely accepted and used by livestock producers; seed, chemical, fertilizer and equipment dealers; agricultural Extension and Natural Resources Conservation Service personnel; vocational agriculture teachers; and agricultural consultants. Our hope that the book might be useful to students in college forage courses was certainly fulfilled as presently it is used at over 30 colleges and universities. We also hoped that this book might have value outside the southern United States. In this we were not disappointed as orders for the book have come from all over the United States and more than 20 foreign countries. Our wide readership suggests that many of the forage-livestock principles covered in *Southern Forages* have no geographic boundaries.

The new revision represented an opportunity to update some sections, add new information, and clarify some statements. Some chapters have been rewritten and/or expanded, including the popular Appendices where a number of new tables are added. Basically, the Second Edition of *Southern Forages* is the same book, only better. We hope that it continues to meet the needs of forage producers, students, teachers, and others. We are grateful for the warm reception our book has received and trust that this new edition will serve your needs well.

Don Ball
Carl Hoveland
Garry Lacefield

Contents

Acknowledgements

The authors gratefully acknowledge the extensive in-depth reviews of the manuscript for this book which were provided by: Dr. Henry Amos, Dept. of Animal and Dairy Science, University of Georgia; Dr. Pat Bagley, North Mississippi Research and Extension Center; Dr. Roy Blaser, University Distinguished Professor Emeritus, VPI and State University; Dr. Larry Grabau, Dept. of Agronomy, University of Kentucky; Mr. Curtis Grissom, County Agent-Coordinator, Limestone County, Alabama; and Mr. Ken Rogers, Alabama Natural Resources Conservation Service State Office.

Other helpful suggestions regarding the manuscript were provided by: Dr. Glenn Burton, Research Geneticist, Coastal Plain Experiment Station, Tifton, Georgia; Dr. Jimmy Henning and Dr. Monroe Rasnake (Dept. of Agronomy), Dr. Roy Burris (Dept. of Animal Science), and Dr. Duane Miksch (Dept. of Veterinary Science), University of Kentucky; Dr. Mark McCann (Dept. of Animal Science), Dr. Troy Johnson and Dr. Harold Brown (retired) (Dept. of Crop and Soil Sciences), University of Georgia; Dr. Edgar Cabrera, Seed Technology Laboratory, Mississippi State University; Dr. Lowell Moser, Department of Agronomy, University of Nebraska; Dr. Lee Stribling (Dept. of Zoology and Wildlife), Dr. David Bransby and Dr. John Everest (Dept. of Agronomy and Soils), Auburn University; Dr. Henry Fribourg, Plant and Soil Science Department, University of Tennessee; Dr. Gerald Evers, Texas A&M University, Overton Research Station; Dr. Larry Murphy, Dr. Bill Griffith and Dr. Bob Thompson, Potash & Phosphate Institute; Dr. Ron Haaland, The Haaland Company, Auburn, Alabama; Dr. Stewart Smith, LiphaTech Inc., Milwaukee, Wisconsin; Mr. William Green and Mr. George Kelley, County Agents in Graves and Hopkins counties, respectively, in Kentucky; and Mr. Ken Johnson, Natural Resources Conservation Service, Monroe County, Kentucky.

Special thanks go to Dr. Thomas A. Powe, Jr., College of Veterinary Medicine, Auburn University, for consultation regarding poisonous plants. The typing of Ms. Thyrza Smith, Ms. Lynn Tidwell, Ms. Carolyn Hightower, and Ms. Christi Forsythe is also gratefully acknowledged. The many contributions of Dr. Bob Darst, Mr. Bill Agerton, and Ms. Kathy Hefner, Potash & Phosphate Institute, and Mr. Charles Hamilton (graphic design consultant), were invaluable. Finally, the authors especially wish to recognize the truly outstanding and untiring work of the editor, Mr. Don Armstrong, also with the Potash & Phosphate Institute.

Photo Credits: The cover photographs of dairy cattle and sheep were provided by Mr. Warren Thompson, ABI Alfalfa, Lexington, Kentucky, and the deer photograph was supplied by Mr. Neil Waer, Dept. of Zoology/Wildlife, Auburn University. The forage crop seed shown in Chapters 5 through 8 were photographed by Mr. Bill Agerton. Other photographs, acknowledged at various points throughout the book were provided by: Dr. D. E. Akin, Dr. G.W. Evers, Mr. Morris Gillespie, Dr. Tom Powe, Dr. Mike Patterson, Dr. Monroe Rasnake, Dr. Ron Shumack, Mr. Warren Thompson, Mr. Neil Waer, and by the National Wild Turkey Federation. Photographs not acknowledged were supplied by the authors. ■

Chapter 1

History of Southern Forage Crops

PRIOR to European settlement, the Southern USA was mainly forest interspersed with small patches of burned areas where Indian women grew crops of corn, beans, and squash. Grazing animals consisted mainly of deer and bison which were hunted for meat (**Figure 1.1**).

During the late 1600s, some Indians began to raise cattle which they acquired from European colonists. There was some grazing of native grasses, but no hay was harvested. Later, Indians such as the Cherokee tribe adopted the European practice of raising cattle, sheep, and horses on planted pastures.

European settlers in the region found big bluestem, indiangrass, and switchgrass growing on favorable soils. These grasses were generally nutritious and relished by livestock, but overgrazing soon reduced vigor and stands, probably resulting in replacement by less desirable grasses such as broomsedge and three-awn grasses. Most of the native grasses on leached, infertile soils common to much of the Southern USA had handicaps such as low nutritive quality, low yield, short productive season, or intolerance to grazing. Thus, animal agriculture based on native grasses was at a distinct disadvantage in the South.

Forage production with improved species developed differently in the South than it did in other parts of the nation. In the Northeast and Midwest, commercial dairy, beef,

Figure 1.1. Bison such as these originally grazed the native grasses of the South.

and sheep production on European-introduced forage plants increased rapidly during the 1800s to supply the needs of large industrial cities and for export to Europe. Natural grasslands in the more arid West were exploited for beef cattle and sheep. In contrast, agriculture in the South consisted of cash crops such as cotton, with livestock being grown on only a limited scale to supply the local demands of small cities and rural areas.

Improved pastures and large-scale beef and dairy production did not occur in much of the South until the 1930s and 1940s. Since then, there has been a massive expansion of cattle production throughout the South and forages have become an important agricultural commodity. Currently, there are over 60 million acres of perennial pasture in the South, making up about 75 percent of the total pasture acreage in the humid Eastern USA. In addition, there are about 19 million acres of annual pastures in the South.

European settlers in the Northeast introduced cool season species such as orchardgrass, perennial ryegrass, timothy, bluegrass, and white clover, which were productive there. As the settlers migrated west and south, they found these species were also adapted to the upper South. Farther south, they succeeded with these cool season forages in the mountains, but failed at lower elevations. Colonists at Savannah, Georgia, introduced alfalfa in 1736, but it failed. Virginia farmers were also disappointed with alfalfa, which performed poorly due to soil acidity and low fertility.

None of the improved forage species commonly grown in the South

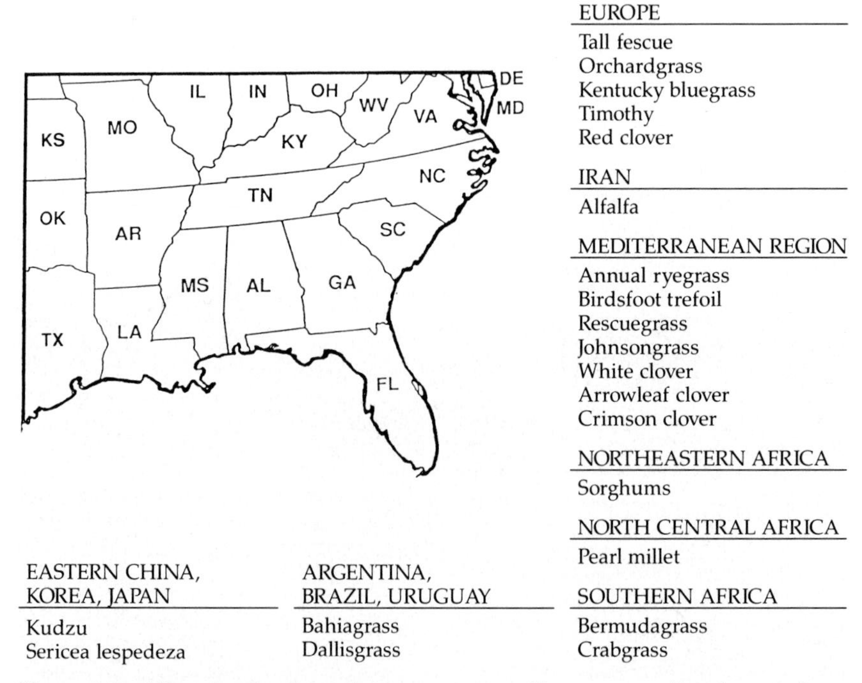

Figure 1.2. All major Southern forage grasses and legumes are immigrants from other parts of the world.

are native to the region or even to other areas of the USA. Most Southern forage plants, like most of the nation's human population, were originally immigrants from other parts of the world (**Figure 1.2**).

Bermudagrass was introduced from Africa, probably in the late 1600s, and soon became one of the most important grasses in the lower South. Dallisgrass, a native of Argentina and Brazil, was introduced accidentally through the port of New Orleans and grown in Louisiana in the early 1800s. Johnsongrass, native to the Mediterranean area, was introduced to South Carolina in the 1830s. In about 1840 it was planted by Colonel William Johnson on the black soils of central Alabama where it became a popular forage grass.

Common bahiagrass was first introduced from Brazil by the University of Florida in 1913. It was found to be well adapted but low-yielding. E.H. Finlayson, an agricultural Extension agent, found a much more vigorous and productive bahiagrass growing near the ship docks at Pensacola, Florida, in 1935. It was eventually released by the University of Florida and the Soil Conservation Service as the variety Pensacola, which now comprises the majority of the bahiagrass acreage in the USA.

Tall fescue, the most widely grown cultivated pasture grass in the USA, probably was originally introduced as a contaminant in other grass seed from Europe. It was found growing on a hill pasture in eastern Kentucky in 1931 by E.N. Fergus, a professor at the University of Kentucky. Tall fescue was growing on this farm when it was purchased by William M. Suiter in 1887 (**Figure 1.3**). Tests revealed that this was a productive grass, and it was released in 1943 as the variety Kentucky 31. There was phenomenal acceptance and planting of this variety in the late 1940s and 1950s, as it was useful in a large portion of the lower Midwest and upper South

Figure 1.3. Kentucky 31 tall fescue originated from plants growing in this pasture on the William Suiter farm, Menifee County, Kentucky.

where other cool season perennial grasses were not well adapted. Later releases of Claire timothy and Boone orchardgrass by the University of Kentucky provided additional forage grasses for the upper South.

The cereal grasses–oats, wheat, and rye–were grown as grain crops in the South from the time of the first European settlements. Later, these grasses became important as high quality winter annual grazing crops. Annual ryegrass, native to the Mediterranean area, was also introduced into the South during the early years of European settlement, and is now the most common component in winter annual pastures.

There was little selection or breeding of forage crops in the South until the 1930s. When G.W. Burton came to Tifton, Georgia, as a forage grass breeder in 1936, common bermudagrass was generally considered a weed in row crops. His remarkable breeding program soon resulted in release of the hybrid variety Coastal which was more drought-tolerant, frost-tolerant, and much higher yielding than common bermudagrass. At present, it is estimated that there are over 12 million acres of this variety in the South.

There have since been many notable achievements in developing improved forage varieties for the South. Dr. Burton's breeding program has released several additional improved bermuda hybrids that are more cold hardy or more digestible and result in higher animal production. Other breeding work at Tifton, Georgia, by Dr. Burton and W.W. Hanna greatly improved the nutritive quality of pearl millet, a summer annual grass native to Africa. Another breeding achievement at Auburn University in Alabama resulted in the release of AU Triumph, the first winter-productive tall fescue variety for the lower South. Several other tall fescue varieties, including Johnstone and Kenhy developed by R.C. Buckner at the University of Kentucky, are also now available.

Legumes have been less successful in the South, a result of low soil fertility, subsoil acidity, heat, drought, pest problems, and generally higher management requirements than grasses. Crimson clover, a winter annual from the Mediterranean region, was introduced into the USA in 1819 but did not become important until 1880. The long-term breeding work of W.E. Knight in Mississippi did much to improve this clover and resulted in release of the winter-productive variety Tibbee. Arrowleaf clover, a winter annual from Italy, was released in the early 1960s. In particular, the Auburn University-developed variety Yuchi became important as a long-season clover in the lower South.

A significant step in the acceptance and use of perennial clovers occurred in the late 1940s when E.N. Fergus released the variety Kenland red clover from the University of Kentucky. Continued red clover breeding in Kentucky by N.L. Taylor resulted in additional disease-resistant varieties for the South. The ladino white clover Regal, developed at Auburn University by P.B. Gibson, E.D. Donnelly, and W.C. Johnson, and the intermediate white clover Louisiana S-1, released by C.R. Owen at Louisiana State University, have both made important contributions to forage/livestock production.

Alfalfa, originally from Iran, was of little importance in the South

Figure 1.4. Tall fescue-white clover-birdsfoot trefoil provides excellent pasture for beef cows and calves. (Tim and Peg Taylor farm in eastern Kentucky.)

until the development of better-adapted, improved varieties and enhanced soil fertilization technology. Alfalfa acreage slowly increased, especially in the upper South, for several decades but fell sharply to a very low level in the 1960s with the arrival of the alfalfa weevil. With weevil control, alfalfa acreage began rebuilding and there are presently more than a million acres in the region.

Common annual (striate) lespedeza was introduced to Georgia from eastern Asia in 1846, later spreading throughout the region. Korean annual lespedeza was introduced in 1920 and became important in the upper South. Sericea, a perennial lespedeza, was introduced from Japan in the late 1890s but did not become important until the 1920s and 1930s. Sericea is especially tolerant of low fertility and acid subsoils, but was considered a low-quality forage plant. Later, the breeding program of E.D. Donnelly at Auburn University developed fine-stemmed and low-tannin sericea varieties with improved palatability and nutritive quality.

Several other forage species have shown promise and some are being utilized. Berseem, subterranean, and rose clovers are valuable winter annuals in certain areas of the lower South. In portions of the Gulf Coast area and in peninsular Florida, perennial peanut is a valuable high-quality hay plant. In the upper South, there is greater use of warm season perennial grasses such as bermudagrass, big bluestem, caucasian bluestem, switchgrass, and indiangrass.

Introduction of improved forage plants, together with advances in fertilization and weed control, has greatly increased productivity of pastures and hay areas. Pasture renovation which has included establishing legumes in grass sods has improved yield and quality of pastures and hayfields. Along with forage improvement, there have been great advances in animal breeding, control of animal breeding season, and animal health care.

Grazing research by scientists such as Roy Blaser in Virginia and G.C. Mott in Florida has shown the great potential of well-managed pasture for livestock production in the South. Probably the least progress has been made by Southern livestock producers in managing pastures and hay fields for high forage quality. Improper grazing often results in surplus grass of reduced quality, and the resultant shading prevents development of new tillers and leaves. Most hay is cut when it is too mature, reducing digestible energy and protein.

Forage production in the South has come a long way, mainly through improved plants, farm equipment inventions, and fertilization. **Table 1.1** lists some significant dates.

The primary challenges for the future are better management of existing grasses and legumes for improved nutritive quality, and matching animal nutrient requirements with the available forages. Adoption of these skills will result in continued progress in the efficiency and profitability of Southern forage/livestock systems. ■

Table 1.1. Some significant events affecting Southern forage/livestock production.

1701	**Grain drill invented by Jethro Tull.**
1733	**J.E. Oglethorpe established an experimental garden for foreign plant introductions at Savannah, Georgia, USA.**
1833	**Obed Hussey developed the first successful reciprocating cutter bar hay mower.**
1837	**John Deere produced the first steel moldboard plow.**
1840	**Wire-toothed hay dump rake developed.**
1840	**Justus von Liebig in Germany founded the science of modern plant nutrition.**
1841	**Edmund Ruffin in Virginia increased the yield of crops such as alfalfa with lime.**
1843	**First fertilizer research on grassland began at Rothamsted, England.**
1850	**Invention of corn planter.**
1862	**Morrill Act created the Land Grant University System in USA.**
1866	**Gregor Mendel in Bohemia hybridized peas and laid the basis for modern genetics and breeding.**
1870	**Disk harrow first produced.**
1872	**First horizontal continual action hay press.**
1873	**First silo constructed in USA.**
1874	**Barbed wire patented for fence.**
1886	**Hellriegel and Wilfarth in Germany discovered that legumes fix nitrogen in nodules of roots.**
1887	**Hatch Act approved, providing funding of USA Agricultural Experiment Station system.**

1890 Side delivery hay rake and hay loader developed.

1903 C.W. Hart and C.H. Parr began manufacture of first gasoline-powered tractors in Iowa.

1908 G.H. Shull discovered the concept of hybrid vigor in corn.

1914 Smith-Lever Act established the Cooperative Extension Service in USA.

1918 D.F. Jones developed the double-cross procedure, making production of hybrid corn seed practical.

1919 First grassland plant breeding station in the world started by R.G. Stapledon at Aberystwyth, Wales.

1928 Future Farmers of America formed in Kansas City, Missouri.

1933 Tennessee Valley Authority approved by Congress.

1933 Agricultural Adjustment Act created a county committee system which became the Agricultural Stabilization and Conservation Service in 1954.

1934 The first Alfalfa Improvement Conference was held in Lincoln, Nebraska.

1935 Soil Conservation Act created the Soil Conservation Service, now called the Natural Resources Conservation Service.

1940 Self-tying pickup hay baler developed.

1940 Southern Pasture and Forage Crop Improvement Association founded, Tifton, Georgia.

1944 Joint Committee on Grassland Farming organized at Rutgers University in New Jersey, becoming the American Grassland Council in 1957 and the American Forage and Grassland Council in 1968.

1945 Introduction and use of anhydrous ammonia as a fertilizer.

1948 Society for Range Management organized in Salt Lake City, Utah.

1953 Certified Alfalfa Seed Council organized.

1958 Development of hay baler for large round bales.

1960s Development of no-till drill for seeding in grass sods.

1962 National Alfalfa Variety Review Board established.

1975 Development of high-voltage electric fence in New Zealand.

1976 Near infrared reflectance spectroscopy described as a method for predicting forage quality.

1977 An on-farm observation by USDA scientists in Georgia indicated an association between a fungal endophyte in tall fescue and fescue toxicity in cattle.

1980s Research allows development of an understanding of the effects of the tall fescue endophyte on animal performance and on the stress tolerance of tall fescue plants.

1990 Release of the first grazing-tolerant alfalfa variety adapted in the South.

1994 First National Alfalfa Grazing Conference held in Nashville, Tennessee.

Chapter 2

Developing a Forage Program

WITHIN a particular community or area of the South, the basic forage production resources may be quite similar on many farms. Yet, there are vast differences in the levels of production, efficiency, and profitability of forage/livestock programs.

Actually, many factors influence the success of a forage/livestock operation, and the approaches taken in developing a forage program rank high on the list. In turn, the approaches taken are influenced by the concepts and attitudes of the producer.

There is no one plan or program that fits all forage production situations. Prior to making forage plantings, many factors need to be considered, including: available soil types; topography; other farm enterprises; availability of capital; labor resources; type of livestock operation; and objectives of the producer. Consequently, there is no one "right" or "wrong" forage program. There are numerous options available and the producer must make careful selections.

Forage/livestock production is complex. Thus, development of a good program requires thought and planning. In fact, a written plan is best because it helps clarify objectives, identify problems or limiting factors, and define needs and opportunities.

Without question, a producer's overall forage production philosophy has an important influence on the type of forage program developed. This chapter introduces several

Figure 2.1. Concepts and attitudes of the producer are critically important in forage/livestock production.

concepts or characteristics which outstanding grasslanders typically embrace or exhibit. These might be considered "basic" concepts, and certainly they deserve consideration by any producer who wants to develop and maintain a good forage program. Several are discussed in greater detail in later chapters.

The Basic Commodity Is Forage

The USA has a livestock industry of which we can justly be proud, but some parts of the world outproduce us, and often do so more efficiently. Climate may explain part of this, but management philosophy is a major difference often noted between forage-based livestock farms in the USA and the highly efficient livestock farms in New Zealand, Europe, and other parts of the world.

This difference in philosophy is that many producers in the USA see themselves basically as **livestock managers** who run their animals on pasture; in other parts of the world where there is outstanding forage/livestock production, the operators see themselves as **grass farmers** who are using livestock to harvest their grass. This subtle difference in philosophy likely accounts for a substantial portion of the success of these competing areas.

While good animal management and breeding are of critical importance, it appears that there are far more American producers who have excellent animal management skills and outstanding breeding stock, but who have poor forage management knowledge, than vice versa. In most cases, profitable livestock production is largely dependent on a producer's ability to efficiently convert forage (hay, pasture, and/or silage) into meat, wool, or milk.

Inadequate digestible energy in the diet is the primary limiting factor in most Southern livestock operations. The solution to this problem on most farms is development of a forage program which increases the digestible energy content of consumed forages.

Use of Reliable Information

It is not unusual for an inexperienced, but highly motivated person to start a cattle or sheep operation and quickly develop a more efficient and profitable business than more experienced neighbors. Likewise, some horse producers or wildlife enthusiasts who are inexperienced with making forage plantings have great success while others, perhaps more experienced, do not.

Such occurrences are not "beginner's luck"; they result from the enthusiastic application of technology. Though initially not knowledgeable, such producers find and utilize sources of reliable information.

Producers need to pay attention to research-based recommendations, particularly university recommendations. Agriculturally-oriented universities are **in the business** of doing scientific research, and have the facilities and trained personnel. Universities are therefore an excellent source of unbiased information.

Great care and precision are required to accurately evaluate forage crops. There are variables which enter into the evaluation of forage crops which are not relevant to other crop species. For example, total pasture yield may be less important than seasonal production and forage quality. In animal feeding, the emphasis should be on the total pounds of available nutrients produced and the economics of animal performance obtained from those nutrients, rather than on forage yield.

It is especially difficult to accurately determine animal performance on pasture through observation of on-farm situations. Livestock prefer certain forage species or varieties, but these may not be the best nutritionally. Differences in palatability may be associated with higher forage quality of the preferred crop, but not always. Grazing animals may perform equally well on a less-preferred forage when it is all that is available. Slight differences among fields with regard to soil type, slope, drainage, or fertility can also affect forage growth and animal preference.

The Cooperative Extension Service is the organization primarily responsible for disseminating research-based agricultural information from land grant universities. This is done in many ways, including local meetings, publications, and through mass media. Extension personnel can also visit individual farms, evaluate an operation, and suggest specific practices which might be implemented to advantage.

Local Extension offices should thus be a prime source of information, but other excellent sources also exist. The Natural Resources Conservation Service, Vocational Agricultural departments, and Agricultural Stabilization and Conservation Service offices offer agricultural information and advice which can be of great help to forage producers. Some non-land grant universities also have faculty who provide agricultural information to producers as a service function.

Much of this type of information is localized, and therefore particularly useful. For example, the general recommendations set forth in this book are subject to alteration depending on local conditions. Local agricultural workers should be able to adapt research findings or other information to practical on-farm situations suitable to the financial and management skills of individual producers.

Industry is another important source of information. As agricultural technology becomes more complex, it is increasingly difficult for universities to work in all promising research areas. Many seed, chemical, and other companies have research departments with highly trained and competent personnel. Their contributions are immense and valuable.

Regardless of the source of research data, it is rare that a particular new technology or product is applicable on every farm. The best approach is one of caution. It is best to first try a new approach or product on a small scale, then expand its use once it has been proven beneficial.

Timeliness

Some producers develop superb forage programs while others of similar intelligence, resources, and experience have little success. It appears that in many cases, the ability to get things done **when they need to be done** accounts for the difference.

Timely action can be important in virtually any aspect of forage production, but is especially critical in forage crop establishment. For example, different crops have differing optimum planting times, often a period of only a few weeks. Weather and soil conditions sometimes interfere with timely planting, but often late planting is simply the result of a producer "not being ready" at the right time. This is particularly unfortunate because missing the optimum planting time often results in a poor or failed stand. Timely planting costs no more, but the return on investment can be much greater.

Timely action can also make a big difference in productivity once a forage crop is established. Timely fertilization can influence forage yield and quality. Timely herbicide application can greatly reduce weed populations. Most importantly, timely adjustment of pasture stocking rates or timely cutting of a crop for hay or silage can make a great difference in forage quality, stand composition, forage utilization, and animal performance.

Eliminating procrastination is essentially a matter of setting priorities. Producers who want a good forage program which will improve their livestock production, efficiency, and profits, need to place a high priority on timeliness.

Use of Adapted Species and Varieties

This is an elementary point, yet many forage plantings fail or perform poorly simply because the crop is not adapted to the site or area. Forage crop adaptation is limited by soil type, climate, and drainage. Therefore, it is important to consider all of the major forage species which are adapted for each field (see Chapters 5 through 8).

A soils map, available free of charge from the Natural Resources Conservation Service in most areas, is a valuable tool in selecting forage species for a particular site. Observing the forage plants which are already growing successfully or volunteering in an area, as well as consulting with persons who have experience and knowledge regarding matching forages to particular fields may also be helpful. These approaches can help a producer understand potential forage options on a field-by-field basis.

Matching Crops to Needs

Once the forage options for each field have been determined, the best choice(s) for each field should be made in accordance with the objectives of the producer. One consideration should be the nutritional **needs** of the species and classes of livestock which are to be grazed or fed forage harvested from each field. For example, a dairy producer would want to plant high quality forage crops in fields which are to provide feed for lactating animals. Lower-quality species might be suitable in areas which are to be used for dry cows. Thus, it is helpful to have some understanding of the forage quality of various species and to choose those which will best meet animal requirements (see Chapters 16 and 17).

Other needs should also be considered. For example, a horse producer might want to plant alfalfa in one field to provide high quality hay, and grow orchardgrass and white clover in another for pasture. In a holding lot where there will be a great deal of trampling, common bermudagrass or tall fescue might be the best choice. Minimizing erosion is always important, but especially in deciding the location of annual pastures which will require frequent soil tillage. Soil survey information can be helpful in planning (**Figure 2.2**).

Maximizing the Length of the Grazing Season

A basic goal of any grazing program should be to grow a variety of forage crops which will provide good grazing **over an extended period of time**, thus reducing stored feed requirements. In general, grazed forages cost only about one-half as much as stored feed due to the costs associated with hay or silage production and feeding.

Using mixtures of forage species can help accomplish this goal. For example, the overall grazing season of a rye-ryegrass mixture will be longer than that of a pasture of small grain alone. The same is true for a red clover-tall fescue or birdsfoot trefoil-tall fescue pasture compared to a pasture consisting only of tall fescue. Any forage mixture should consist of species which are adapted to the field in question, and which are compatible with each other.

A second approach is to plant various forage species in different fields and allow the animals access to each field only when it is making productive growth. When possible, it is advantageous to plant some fields to warm season species and other fields to cool season species. For example, in some parts of the South, it is possible to have on the same farm some acreage of cool season pasture species such as white clover and tall fescue and some acreage of warm season species such as bahiagrass, dallisgrass, bermudagrass, and/or sericea lespedeza.

In the lower South, overseeding winter annuals on dormant sods of warm season species such as bahiagrass or bermudagrass is an excellent way to extend the grazing season (see Chapter 12). In addition, autumn tall fescue growth can be "stockpiled" by deferring grazing until cold weather, thus delaying the feeding of hay. Also, if grain crops are grown on the farm, it may be feasible to turn animals into harvested fields to allow the animals to glean residues.

It is less expensive and less trouble to allow **animals** to harvest the forage whenever possible. This philosophy should be reflected in the planting and management decisions made for different fields on a livestock farm.

Soil Testing and Fertility

Proper fertilization and liming normally result in more dramatic increases in forage production than any other single practice. Thus, it is a management tool which deserves special attention.

It is highly desirable to keep a record of the soil fertility and pH status of each field. One good approach is to make a file for each field in which to keep soil test reports as well as a record of lime and fertilizer applications.

The use of a field activity record can be quite helpful. A sample form is presented in **Figure 2.3,** later in this chapter. This form can be altered to include other types of information such as stocking rates, and dates grazing was begun or ended.

Fertilizing and liming at the proper times and rates recommended by soil tests are basic to good pasture management. For most forage species, certain minimum levels of fertilization and liming are necessary just to maintain a forage stand. However, heavy fertilization is much more feasible when stocking rates are high than when they are low. Detailed information regarding soil testing, fertilization, and liming is provided in Chapter 9.

Use of Legumes

Growing legumes with grasses may be more profitable than growing grass alone. Although legumes are usually more difficult to manage, their advantages often outweigh their disadvantages. Not every acre of forage crops should include legumes, but it is worthwhile for a producer to consider whether the introduction of legumes would be of benefit and be feasible. Chapters 7, 8, and 13 provide additional information on forage legumes.

Courtesy of Mr. Morris Gillespie

Figure 2.2. In many areas, soil survey documents are available free of charge from the Natural Resources Conservation Service (formerly Soil Conservation Service).

Stored Feed

Even though the South has a long growing season and high rainfall relative to many parts of the USA, most livestock producers will need to make some provisions for stored feed to use during periods when pasture is not available. In developing or evaluating a forage program, factors such as amount and quality of stored feed needed, and convenient location of fields should be considered.

The cost of storing feed for livestock has risen steadily. Therefore, the best interests of livestock

Forage Field Activity Record

CROP GROWN: ____ **NUMBER OF ACRES:** ____ **CROP YEAR:** ____ **FIELD NUMBER/NAME:** ____

Variety or hybrid ____ Seed dealer ____

Seed: Germination % ____ Vigor test ____ Other details ____

Seeder type and setting ____ Seeding rate ____ Depth ____ Planting date ____

Harvest or graze-down dates (1) ____ (2) ____ (3) ____ (4) ____ (5) ____

Field History: Previous crop ____ Planting date ____

Variety ____ Seeding rate ____

Last year's yield ____ Number of cuttings ____

Soil and Tillage Information: Soil type ____

Tillage method: Moldboard Plow ____ Chisel ____ No-till ____ Other ____

Describe primary tillage method ____ Date ____

Describe secondary tillage methods ____ Dates ____

Other comments on soil and tillage:

Fertilization Practices:

Method	Timing		N	P_2O_5	K_2O	Secondary Nutrients	Micronutrients
			pounds applied per acre				
Broadcast	Preplant						
Topdress	Harvest (or graze-down)	1					
		2					
		3					
		4					
		5					

Other (manure, rate) ____

Type of lime ____ Year applied ____ Rate ____ CEC, % ____

Crop Protection Chemicals:

List: Formulation Rate Method of application/incorporation Date/Time/Conditions

Herbicides

Insecticides

Other products

Herbicide history	Last year	Two years ago
Chemicals & formulations:		
Rate per acre:		
Method of application:		
Crop injury (description):		

(over)

Figure 2.3. Forage field activity report; other side of form appears on next page.

producers are served by striving to maximize production in fields devoted to the production of stored feed. This is true with regard to both yield and quality.

Typically, at least 30 percent of the costs associated with hay production are **fixed** costs (i.e. purchase of hay baler, tractor, rake, etc.). These costs will be essentially the same regardless of yield. Therefore, increasing yield is beneficial even if it involves an increase in variable costs (i.e. labor, fuel) by 10 to 20 percent or more. Many producers should carefully consider whether the size of their operation warrants the purchase of hay equipment.

Weeds Present:
Kind ______

Insects Present:
Kind ______

Diseases Noted:
Kind (severity) ______

Other problems noted:
Temperatures, etc. ______

Moisture Conditions:
Seasonal observations for this crop ______

Rainfall (total/year) ______ **Rainfall (3 months-growing season)** ______
Any crusting to reduce water intake? ______
Did crop have times of "wet feet"? ______ When? ______
Any wilting periods during growing season? ______ When? ______

Field Appearance:

Date	Poor	Fair	Good	Excellent

Obvious deficiency (D) or toxicity problems
N ___ P ___ K ___
Ca ___ Mg ___ S ___
B ___ Cu ___ Fe ___ Mn ___ Mo ___ Zn ___
Other ______

Field Tissue Tests: (VL, L, M, H, VH) Plant part ______ Method ______ Date ______
Best area: N ___ P ___ K ___ Other ______
Poor area: N ___ P ___ K ___ Other ______
If additional tests are done, note here and attach a record of results.

Soil Test and Plant Analysis Information:

Soil Laboratory Analysis						Recommendations					
							Pounds/Acre				
Field Identification	pH	P	K	Mg	S	Lime, tons/A	N	P_2O_5	K_2O_5	Mg	S

Have plant analyses been taken from problem areas? ______ When? ______
When were soil samples last taken? ______ Results available? ______

General Observations of Field Conditions:

Many small producers are better off buying hay or having it custom baled.

Usually it is desirable to produce the best quality hay or silage possible. In feeding livestock, it is the pounds of nutrients and not the pounds of feed they receive which determine performance. Chapters 18, 19, and 20 provide additional information regarding the production and utilization of hay and silage.

Grazing Methods

A final consideration in developing or evaluating a forage program is how to best utilize the forage produced. Most Southern livestock producers use a continuous stocking method (continuous grazing– see Chapter 25). While this is most convenient, it may or may not be the most efficient, feasible, or profitable. Fencing, water, and labor costs will normally be higher with rotational stocking (rotational grazing). Production and net returns may or may not be higher, depending on the situation.

When evaluating or planning a forage program, a producer should consider what will be the most suitable and convenient way to utilize pastures. The decisions may influence fencing patterns, fertilization practices, labor requirements, and even species selection.

Summary

Development of a good forage program requires time and thought, but is worth the effort. Having basic concepts and principles in mind throughout the planning process is important. Keeping good written records also greatly facilitates proper management and wise decision making. Good forage programs don't just happen, they are **planned**. ■

Chapter 3

Climate and Soils Areas

CLIMATIC conditions largely determine the adaptation of forage species. From the standpoint of plant survival and growth, the most critical aspects of climate are generally temperature and the distribution and amount of rainfall. Climate extremes are more likely to determine adaptation than are average conditions. While local conditions are of primary concern to individual producers, an assessment of the forage/livestock potential of the region is made easier by taking a broader view.

The Southern USA from eastern Texas and Oklahoma to the Atlantic Coast is described as a humid or high rainfall region, but total rainfall varies substantially in the region. Total annual rainfall increases from 40 inches in northern Kentucky and Virginia to 65 inches along the Gulf Coast of Louisiana, Mississippi, Alabama, and Florida. Rainfall declines from 50 inches in

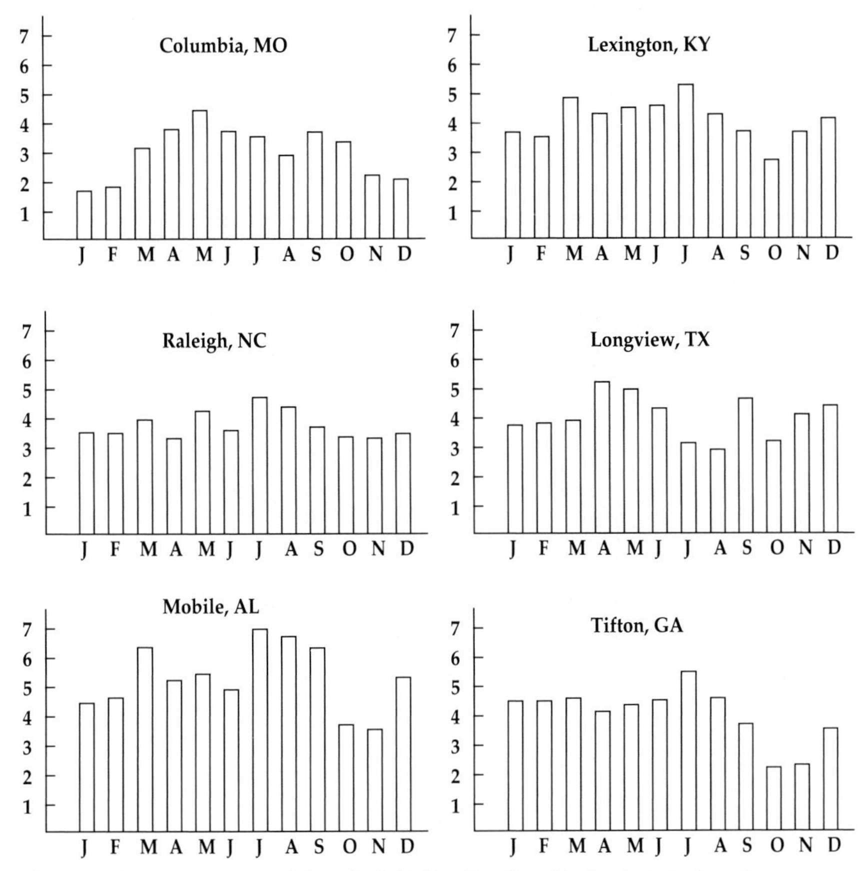

Figure 3.1. Average monthly rainfall distribution (inches) at six locations.

western Louisiana and Arkansas to 40 inches in eastern Oklahoma and Texas.

Generally, rainfall in the South is distributed reasonably well throughout the year (**Figure 3.1**). However, there are periods of the year when rainfall is low or undependable, thus creating problems for forage production in certain parts of the South. In areas such as Missouri, rainfall is highest in spring with occasional droughts in late summer, reducing summer production of cool season perennial grasses. Farther east in Kentucky and North Carolina, rainfall is usually adequate during the year except for a drier period in October when autumn production is reduced.

East Texas normally has good rainfall from autumn through spring, creating excellent conditions for winter annual pastures. However, hot summers with low and poorly distributed rainfall may reduce output of warm season grasses. Adequate rainfall typically occurs all year in the Gulf Coast area (Mobile, Alabama). With mild winter temperatures, forage growth in this area is excellent, but pasture flooding and leaching of soil nutrients may be problems.

Farther east in south Georgia, autumn droughts are common and often make it difficult to establish winter annuals. Soils with low water-holding capacity accentuate the effects of drought. In each case, local rainfall distribution and soils must be taken into account when selecting species, stocking rates, and amount of stored forage.

Summer rainfall at lower elevations in the South originates from convectional storms, often accompanied by thunder and lightning.

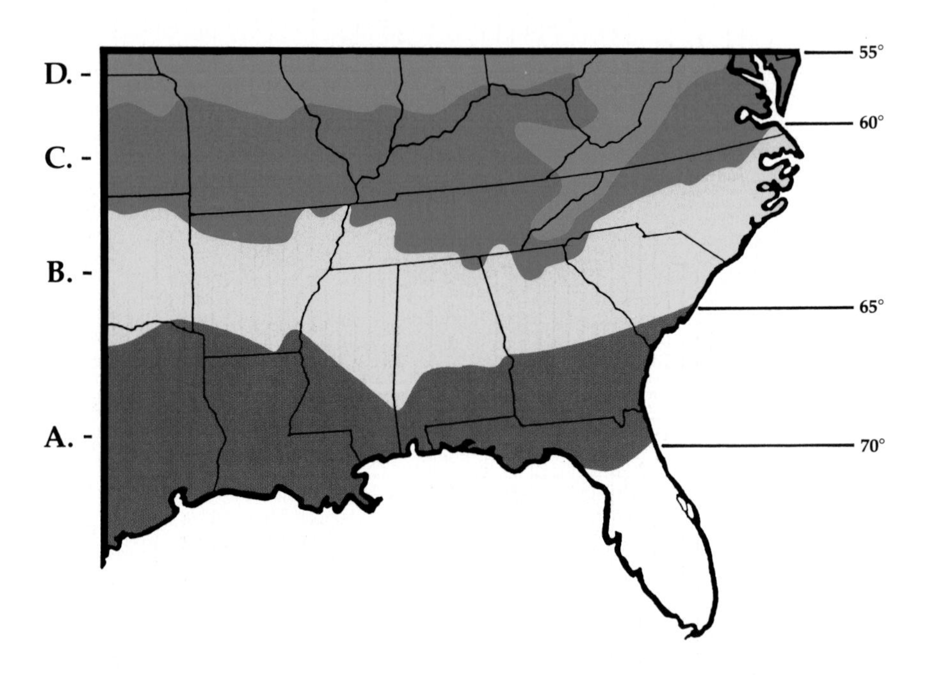

Figure 3.2. Mean annual air temperatures, °F, and adaptation Zones, A, B, C, D.

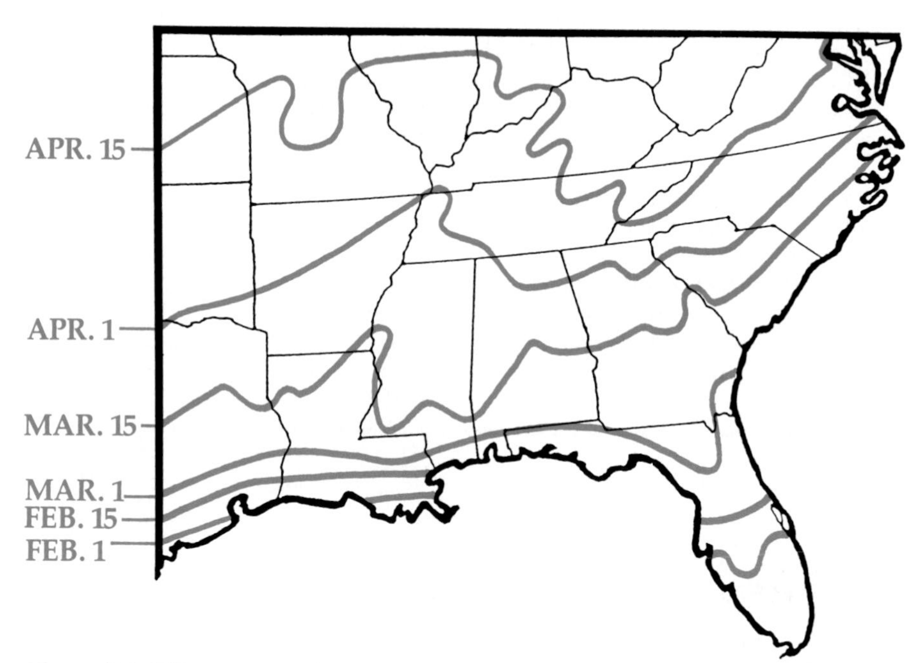

Figure 3.3. Fifty percent chance that temperature will fall to 32°F in spring. Based on average elevation for area. Dates shift as elevation changes.

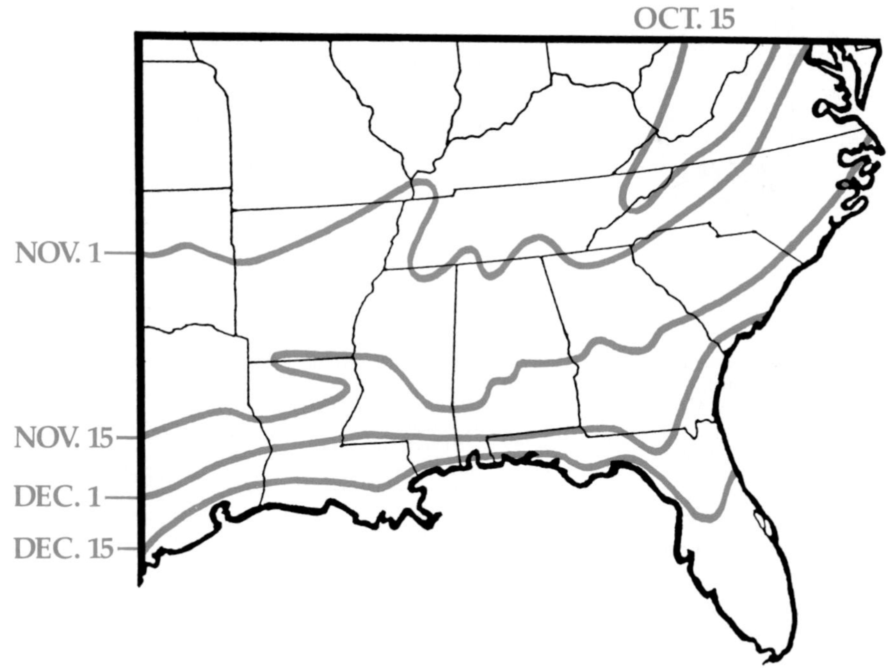

Figure 3.4. Fifty percent chance that temperature will fall to 32°F in autumn. Based on average elevation for area. Dates shift as elevation changes.

The South has more thunderstorms than any other part of the USA. The storms are intense and often of short duration, dropping large amounts of water on the land, exceeding potential soil infiltration rates, and resulting in rapid runoff of water. Thus, although the total amount of measured rainfall is high, the **effective rainfall** available to plants may be low. This problem is even more serious on clay than on sandy soils where infiltration rates are higher.

Days are hot and sultry in summer. Nights are oppressive. The humid atmosphere prevents rapid loss of heat that takes place in the drier climates and clearer air in the western USA. In addition, there is often little air movement as the southern USA is the least windy part of the country. This combination of high summer temperatures and humidity with little air movement favors disease development on susceptible forage varieties. In addition, the summer climate

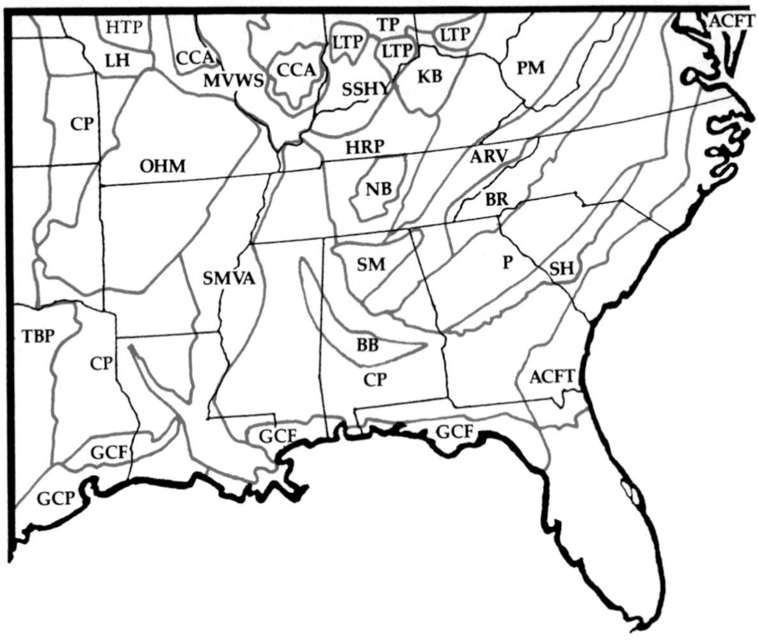

Mississippi Delta Region
- SMVA Southern Mississippi Valley Alluvium

South Atlantic and Gulf Slope
- CP Coastal Plain
- BB Black Belt
- P Piedmont
- SH Sand Hills

Atlantic and Gulf Coast Lowland
- GCP Gulf Coast Prairies
- GCF Gulf Coast Flatwoods
- ACFT Atlantic Coast Flatwoods and Tidewater

East and Central Farming and Forest Region
- OHM Ozark Highlands and Mountains
- SSHV Sandstone and Shale Hills and Valleys
- KB Kentucky Bluegrass
- HRP Highland Rim and Pennyroyal
- NB Nashville Basin
- PM Plateau and Mountains
- ARV Appalachian Ridges and Valleys
- SM Sand Mountain
- BR Blue Ridge

Southwestern Prairies
- TBP Texas Blackland Prairie

Central Feed Grains and Livestock Region
- LH Loess Hills
- HTP Heavy Till Plain
- CP Cherokee Prairies
- CCA Central Claypan Areas
- LTP Loess and Till Plain
- TP Till Plain
- MVWS Mississippi Valley Wooded Slopes

Figure 3.5. Land use (physiographic) areas of the South.

makes haymaking of legumes difficult and favors rapid-drying grasses for hay. Hurricanes during late summer and autumn, although infrequent, can dump large quantities of water over extended areas and cause flooding of pastures.

Winter rainfall is of frontal origin where humid, warm Gulf air meets cool polar air and is forced upward, resulting in persistent cloud cover and rains which may fall steadily for many hours and sometimes several days. The wet cool winters favor growth of cool season forages, especially in the lower South. Snow may fall but it rarely stays on the ground for more than a few days except in the upper South and mountain areas.

Average annual temperatures increase from north to south in the region (**Figure 3.2**). Even though temperatures are generally moderate in the South, the continental effect of the North American landmass exerts a strong effect on winter temperatures. Outbreaks of polar air occasionally cause sharp drops in temperature during winter. These occasional severe cold waves determine survival and adaptation of many forage species. Summers are long and hot, especially in the lower and western part of the region, resulting in the use of warm season perennial grasses followed by cool season annuals. Higher elevations in the Appalachian mountains moderate the summer temperatures, favoring cool season species such as tall fescue, Kentucky bluegrass, orchardgrass, timothy, and clovers.

For the purpose of illustrating the general areas of adaptation of forage species, within this book the region has been divided into four major temperature zones designated as A, B, C, and D (**Figure 3.2**). Moving northward or to higher elevations, the length of the frost-free season decreases with frosts occurring later in spring and earlier in autumn (**Figures 3.3 and 3.4**).

The Southern USA can be divided into different land use areas based on soils, natural vegetation, and agriculture (**Figure 3.5**). Reference will be made to some of these areas in describing adaptation and use of various forage species.

Because of the diversity of soils within the South, it is not feasible to describe in this book the soils in each part of the region. In general, sandy soils predominate in the Coastal Plain while clay, loam, or shaley soils are found farther north. Local descriptions of soils in a particular area can be obtained from the Natural Resources Conservation Service or the Cooperative Extension Service.

Soils often modify the effects of climate, as they differ greatly in water and nutrient-holding capacity, native fertility, and potential for pests such as nematodes. This often has a major effect on adaptation and productivity of particular grasses and legumes.

Generally, soils throughout much of the region tend to be acid and low in nutrients. Acid subsoils which cause aluminum or manganese toxicity are a problem in the Piedmont, Coastal Plain, and Flatwoods areas and limit adaptation of some forage species. Certain soils of chalk or limestone origin in the Alabama-Mississippi Black Belt, Nashville Basin of central Tennessee, Kentucky Bluegrass, and western parts of the South have a near neutral pH, favoring growth of lime-loving legumes. Soil tests are essential to determine lime and fertilizer needs on all soils. ■

Chapter 4

Overview of Southern Forages

MORE THAN 40 species of forage crops are routinely grown in the South, although many can be successfully grown only in certain areas. At least 20 others frequently volunteer or are at least occasionally planted. These crops differ widely in their characteristics, potential uses and value. However, putting them into various categories makes it easier to understand the potentials for using them.

The three criteria normally used to categorize Southern forage crops are whether they are: (l) grasses or legumes; (2) annuals or perennials; and (3) warm season or cool season plants.

Table 4.1 gives a general classification of Southern forage crops. Brief definitions are as follows:

Grass–Botanically speaking, a grass is any member of the *Poaceae* plant family. Grasses are generally herbaceous (not woody), they have parallel leaf veins, fibrous root systems, and bear seed on an elongated seed stalk (see **Figure 4.1**).

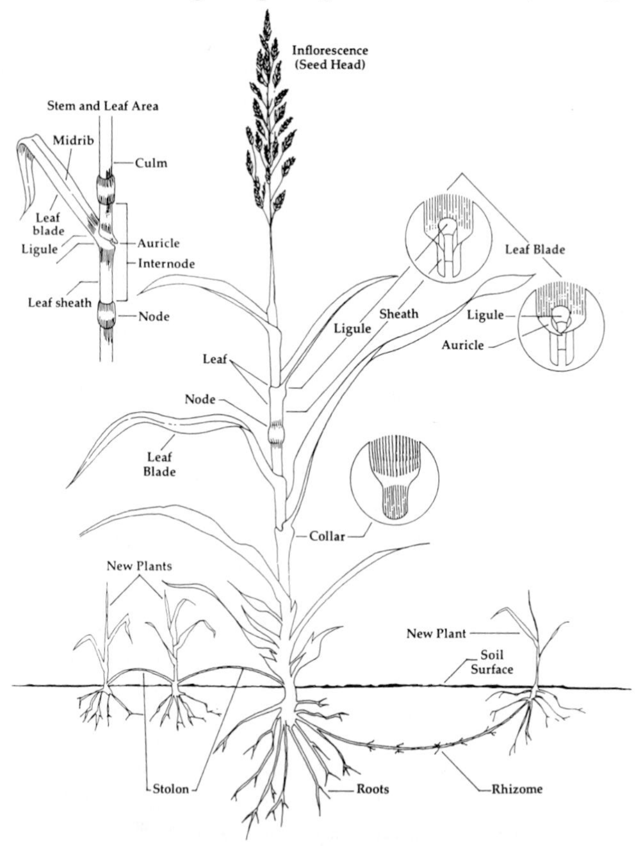

Figure 4.1. Parts of the grass plant. Generalized illustration–not a particular species.

Grasses are monocots (produce only one seed leaf) but should not be confused with sedges and rushes which are also monocots.

Legume–Members of the *Fabaceae* plant family. Legumes are dicots (produce two seed leaves), produce seed in a pod, have "netted" leaf

Table 4.1. A general classification of forage crops frequently grown or found in the South.

GRASSES			
Perennials		Annuals	
Warm Season Perennial Grasses	Cool Season Perennial Grasses	Warm Season Annual Grasses	Cool Season Annual Grasses
Bahiagrass	Kentucky bluegrass	Browntop millet	Barley
Bermudagrass[1]	Orchardgrass	Corn	Oats
Big bluestem	Prairiegrass[2]	Crabgrass	Rescuegrass[2]
Carpetgrass	Reed canarygrass	Forage sorghum	Rye
Caucasian bluestem	Tall fescue	Foxtail millet	Ryegrass
Dallisgrass	Timothy	Grain sorghum	Triticale
Eastern gamagrass		Pearl millet	Wheat
Indiangrass		Sorghum-sudan hybrids	
Johnsongrass		Sudangrass	
Switchgrass			

LEGUMES			
Perennials		Annuals	
Warm Season Perennial Legumes	Cool Season Perennial Legumes	Warm Season Annual Legumes	Cool Season Annual Legumes
Kudzu	Alfalfa	Alyceclover	Arrowleaf clover
Perennial peanut[1]	Alsike clover	Cowpea	Ball clover
Sericea lespedeza	Birdsfoot trefoil	Korean lespedeza	Berseem clover
	Red clover[2]	Soybean	Bigflower vetch
	White clover	Striate lespedeza	Black medic
		Velvetbean	Burclover[3]
			Button clover
			Caleypea
			Common vetch
			Crimson clover
			Hairy vetch
			Hop clover[3]
			Lappa clover
			Persian clover
			Rose clover
			Subterranean clover
			Sweetclover[2,3]
			Winter pea

[1]Frequently or exclusively vegetatively propagated.
[2]This plant may act as an annual or as a biennial–i.e., plants may, under proper management and environmental conditions, live two years before dying.
[3]More than one species is present within the South.

venation, and usually have a taproot type of root system. Most legumes have the ability to interact with bacteria of the genus *Rhizobium* to "fix" nitrogen (N) in nodules on their roots (see **Figure 4.2**).

Annual–A plant which germinates, grows, reproduces, and dies in one year's time or one growing season. Annuals reproduce only by seed.

Perennial–A plant which, under suitable conditions, has the ability to live for more than one year. A perennial may appear to "die back" or become dormant at certain times of the year, but can recover from tubers, rhizomes, or stolons in succeeding years. Perennials may

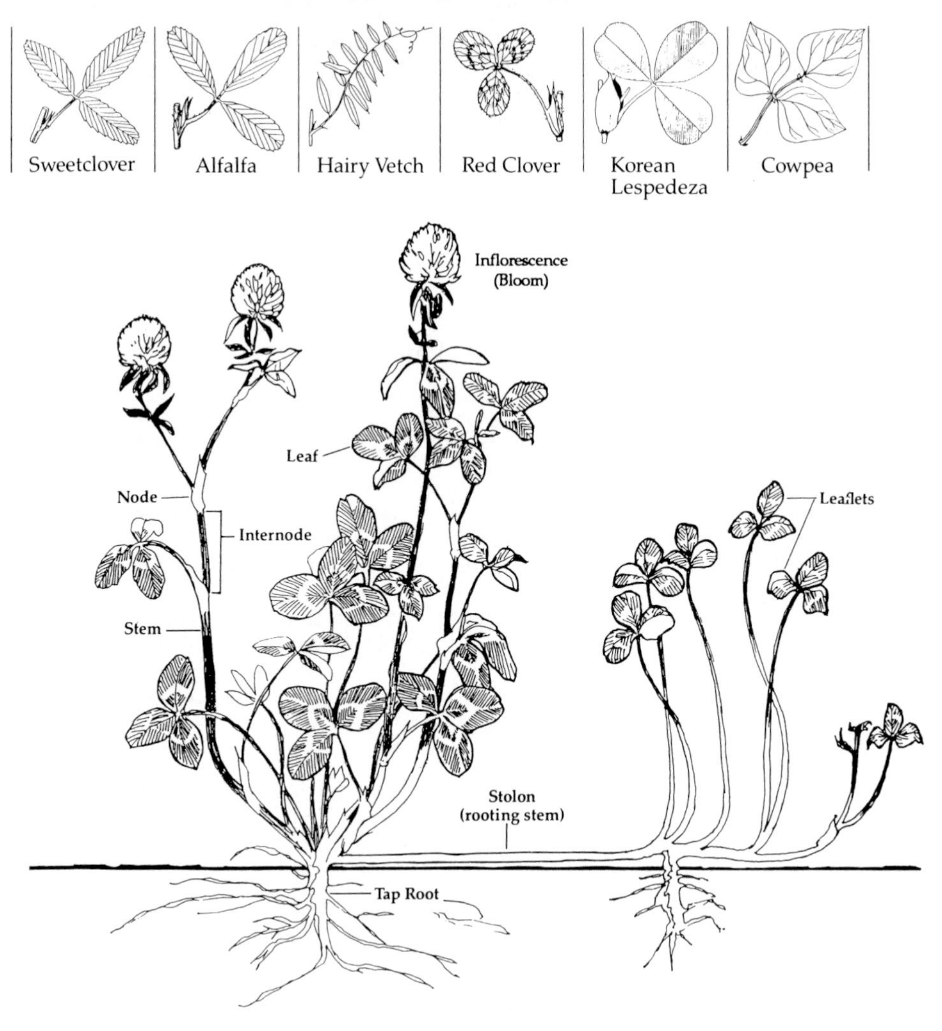

Figure 4.2. Characteristics of some forage legumes. Generalized illustration–not a particular species.

reproduce by seed as well as vegetatively. A few plants are "biennials." These plants are really weak perennials which usually die after the second growing season.

Warm Season Plants–In general, warm season species begin growth and/or are planted in the spring or early summer and make most of their growth during the warmest months of the year.

Cool Season Plants–In general, cool season species begin growth and/or are planted in the autumn or sometimes early spring and make most of their growth during the coolest months of the year, except for the coldest periods during winter.

Distribution of Growth

The distribution of forage growth varies greatly among species and is extremely important in planning a grazing program. Examples of the approximate growing season of a few commonly-used species are shown in **Figures 4.4** through **4.7** for Zones A, B, C, and D (see also **Figure 3.2**).

It should be emphasized that these figures are only rough approximations. The actual growing season will vary with rainfall, timing and amount of fertilization, temperature, variety, and other factors. Furthermore, these figures are less accurate with regard to depicting total forage yield than showing usual seasonal distribution of growth. Therefore, comparison of total growth of various species as indicated by these figures could be misleading. ■

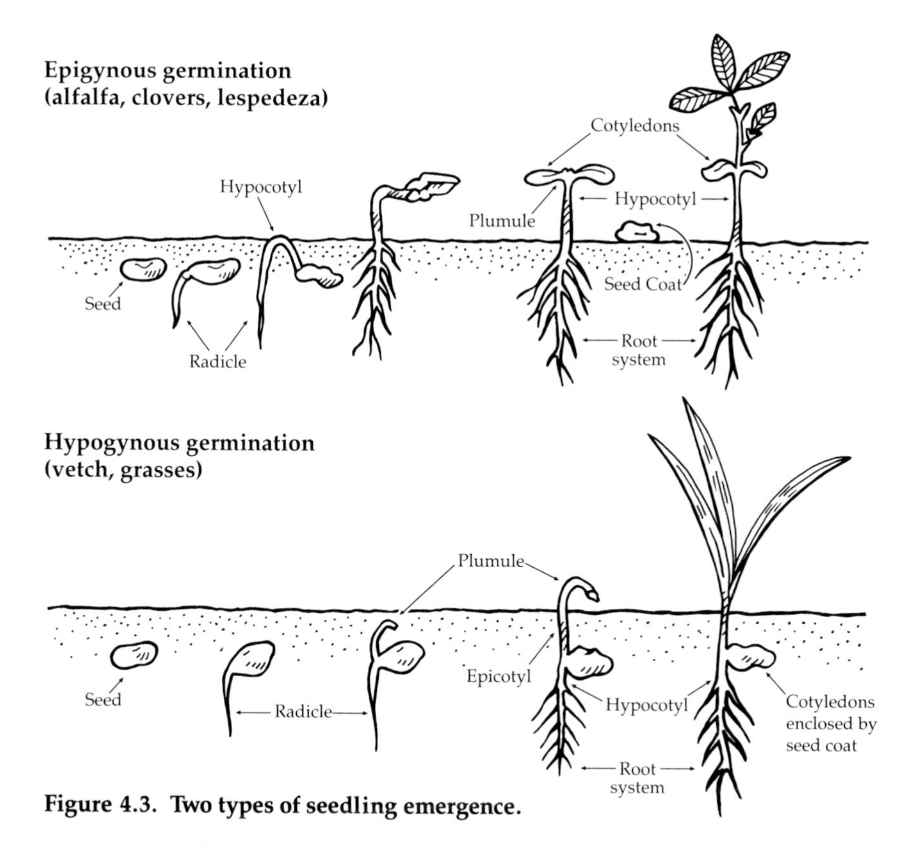

Figure 4.3. Two types of seedling emergence.

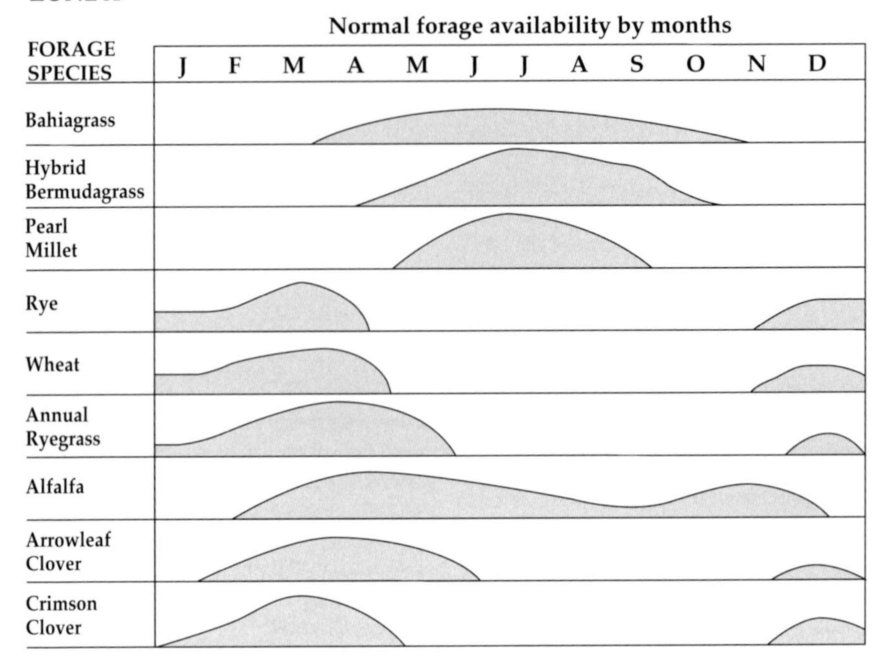

Figure 4.4. Seasonal forage availability of common forage species in Zone A.

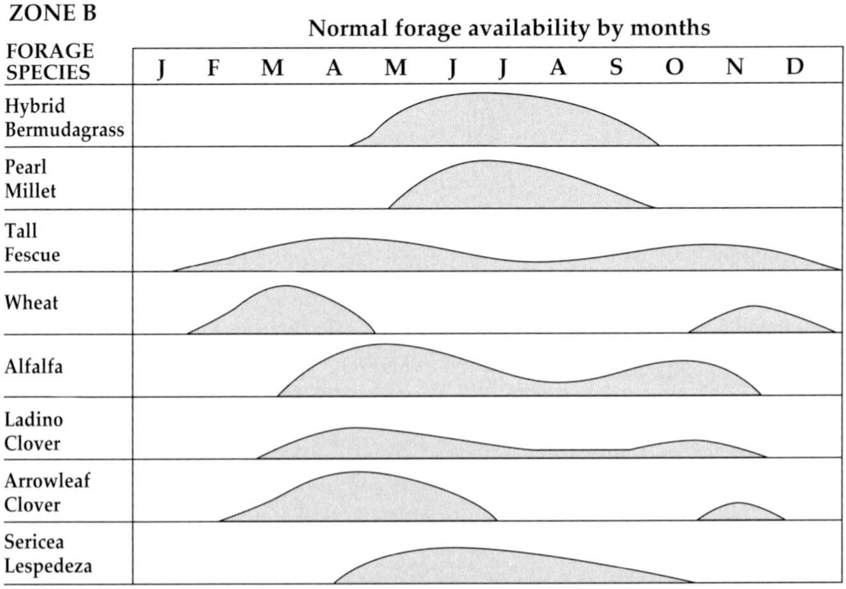

ZONE B

Normal forage availability by months

FORAGE SPECIES

J F M A M J J A S O N D

Hybrid Bermudagrass

Pearl Millet

Tall Fescue

Wheat

Alfalfa

Ladino Clover

Arrowleaf Clover

Sericea Lespedeza

Figure 4.5. Seasonal forage availability of common forage species in Zone B.

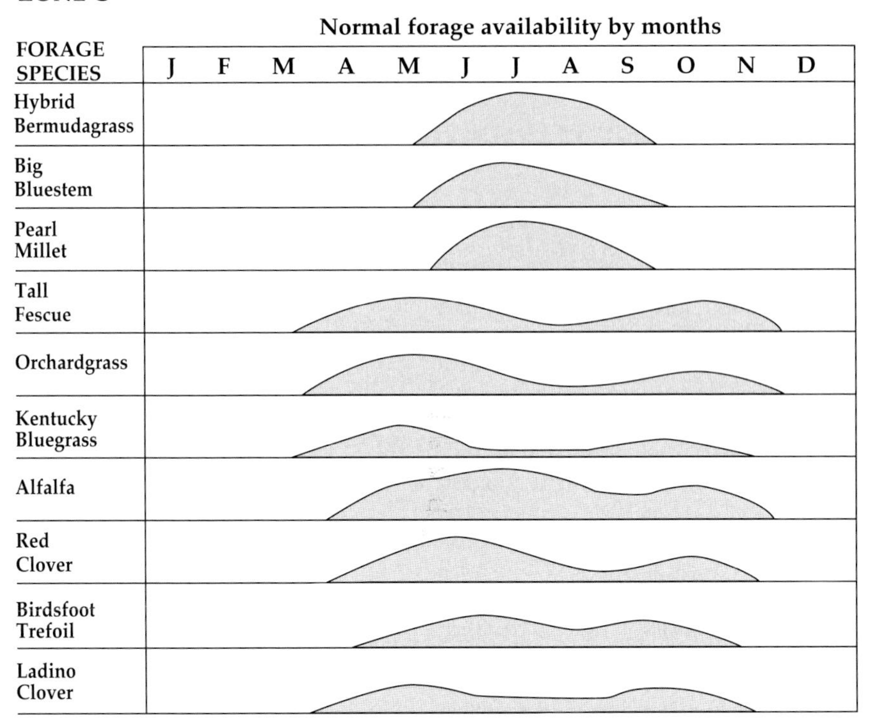

Figure 4.6. Seasonal forage availability of common forage species in Zone C.

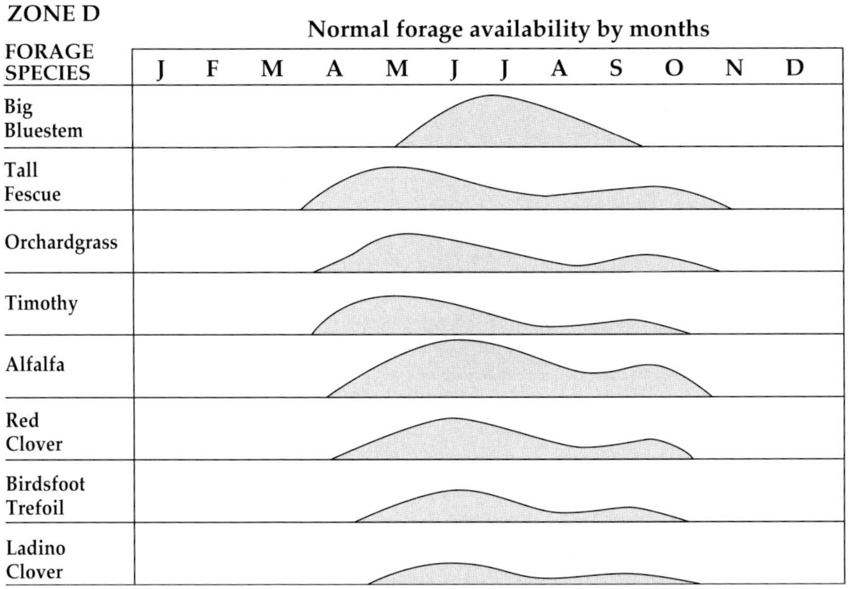

ZONE D

Normal forage availability by months

FORAGE SPECIES

J F M A M J J A S O N D

Big Bluestem

Tall Fescue

Orchardgrass

Timothy

Alfalfa

Red Clover

Birdsfoot Trefoil

Ladino Clover

Figure 4.7. Seasonal forage availability of common forage species in Zone D.

Chapter 5

Warm Season Grasses

WARM SEASON GRASSES grown in the Southern region are of tropical origin and grow mainly during the late spring, summer, and early autumn. Frost kills warm season annual grasses, while warm season perennial grasses become dormant and remain unproductive during the winter months. The productive season of warm season grasses can be as little as 3 months in the northern part of the region (Zones C and D) but may be 7 or 8 months in the Gulf Coast area (lower part of Zone A). Adaptation zones are provided in **Figure 3.2.**

Some of these grasses have the potential for very high forage yields with good fertilization. Unfortunately, forage quality (see Chapter 16) of warm season grasses, especially of perennials, is generally much lower than that of cool season grasses. Plant breeding has resulted in significant improvements in forage quality of some warm season grasses. Winterkilling is a problem with some species or varieties of perennial warm season grasses, especially in northern parts of the region.

Bahiagrass

Bahiagrass *(Paspalum notatum)*

Origin: Argentina, Brazil, Uruguay, Paraguay.

Description: Perennial. Spreads by rhizomes and seed, forms a dense sod. Very aggressive. Deep rooted. Grows 12 to 20 inches tall.

Primary Adaptation: Zone A and lower part of Zone B. Best adapted on sandy soils. Tolerant of drought and poor drainage.

Major Uses: Pasture, hay, erosion control.

Establishment: Seed are planted at 15 to 20 lb/A in March or April.

Fertilization: Very tolerant of low fertility and soil acidity. Responds to nitrogen and potassium.

Seasonal Production: April-October.

Management: Best used for pasture. Close grazing is desirable. Overseed with winter annuals if desired.

Pests: No major pests.

Bermudagrass

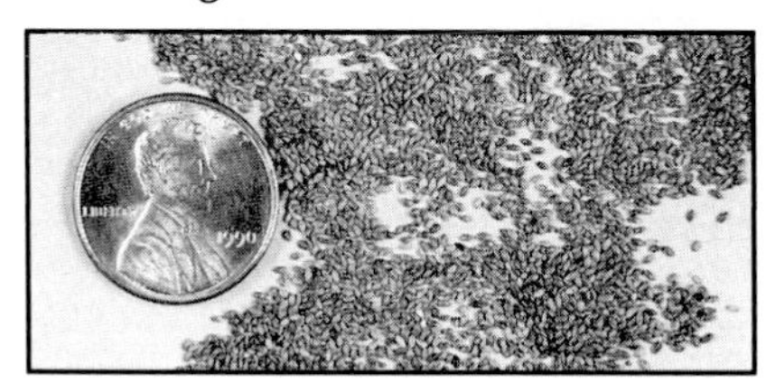

Bermudagrass field scene.

Bermudagrass *(Cynodon dactylon)*

Origin: Southeastern Africa

Description: Perennial. Spreads by rhizomes, stolons, and (in some types) by seed. Hybrids are deep-rooted. Grows 15 to 24 inches tall.

Primary Adaptation: Best in Zone A but certain varieties are cold-hardy in Zones B and C. Best adapted on sandy soils. Extremely drought-tolerant.

Major Uses: Pasture, hay.

Establishment: Hulled seed of common bermudagrass or other seed-propagated varieties or types should be planted at 5 to 10 lb/A in spring. Hybrids are planted in March-April with sprigs at 10 bu/A in rows or broadcast at 25 to 40 bu/A and covered.

Fertilization: Highly responsive to nitrogen. Potassium is important for survival and production.

Seasonal Production: May-October in Zone A, late May-September in Zone B.

Management: Hay should be harvested at about 4 week intervals. With good management, hay yields of 5 to 7 tons/A can be obtained. Should be closely grazed to maintain quality. Annual clovers, small grains, and ryegrass should be overseeded in autumn if winter-spring production is desired.

Pests: Fall armyworm and spittlebug. Leafspot may occur, especially when soil potassium levels are low.

Big Bluestem

Browntop Millet

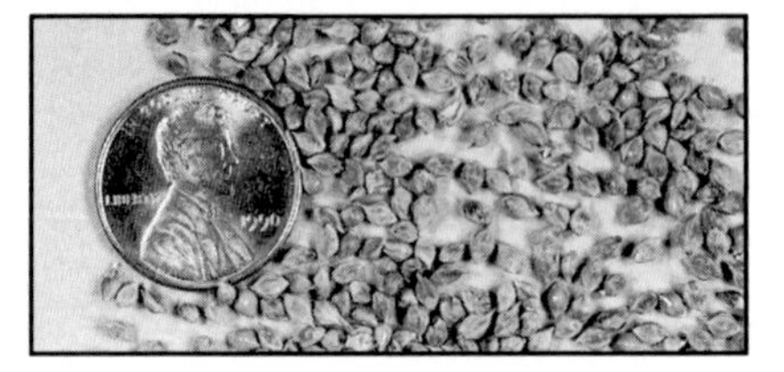

Big Bluestem
(Andropogon gerardii)

Origin: Great Plains and eastern USA.

Description: Perennial bunchgrass, sometimes having rhizomes. Deep-rooted. Grows 3 to 6 feet tall. More drought-tolerant than most warm season grasses.

Primary Adaptation: Western parts of Zones C and D. Little is planted in other parts of the South at present.

Major Uses: Pasture, hay. It remains palatable and nutritious over a longer time than switchgrass.

Establishment: Slow seedling establishment. Seed should be planted at 5 to 10 lb/A pure live seed in April-May.

Fertilization: Responsive to nitrogen.

Seasonal Production: June-August.

Management: Will not tolerate close, continuous stocking.

Pests: Minor.

Browntop Millet
(Panicum ramosum)

Origin: Southeast Asia

Description: Annual. Erect, 2 to 3 feet tall, leafy, fine-stemmed.

Primary Adaptation: Zones A, B, C, and D. Does not produce well on calcareous soils.

Major Uses: Pasture, hay.

Establishment: Seed drilled at 15 to 20 lb/A or broadcast at 25 to 30 lb/A in May-August.

Fertilization: Responsive to nitrogen, very tolerant of soil acidity.

Seasonal Production: June-August. Much less productive than pearl millet or sorghum-sudan hybrids.

Management: Hay should be cut at heading.

Pests. Fall armyworm.

Carpetgrass

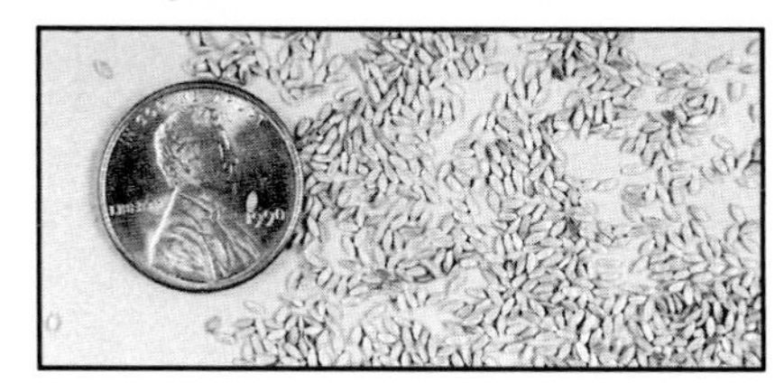

Carpetgrass ***(Axonopus affinis)***

Origin: Central America and West Indies.

Description: Perennial, low-growing, spreading by stolons and seed, forms a dense sod. Grows 6 to 10 inches tall.

Primary Adaptation: Zone A, low soil fertility.

Major Uses: Pasture. Low nutritive quality. Low yield.

Establishment: Seed.

Fertilization: Not very responsive to fertilization.

Seasonal Production: May-August.

Management: Close grazing.

Pests: No major pests.

Crabgrass

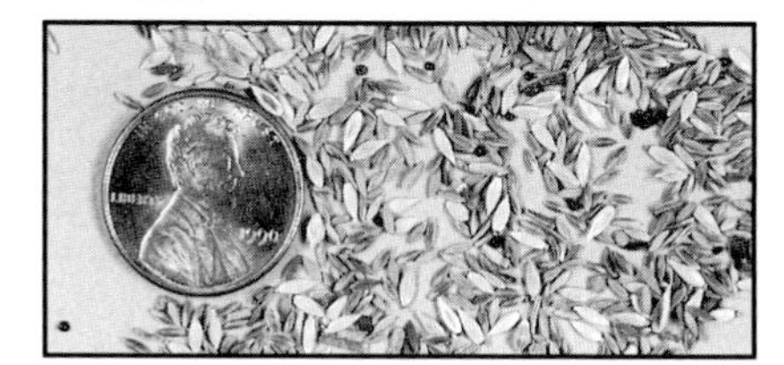

Crabgrass ***(Digitaria sanguinalis)***

Origin: Southern Africa

Description: Annual, creeping-type growth with long runners, very leafy. Grows 2 to 4 feet tall.

Primary Adaptation: Zones A, B, and C. Most productive in the Gulf Coast area with high rainfall.

Major Uses: While considered a weed in many farming situations, crabgrass is useful for pasture and hay. Nutritive quality is superior to warm season perennial grasses.

Establishment: Seed. Reseeds well.

Fertilization: Crabgrass is highly responsive to nitrogen fertilizer.

Seasonal Production: May-September.

Management: With adequate fertilization, crabgrass will support good stocking rates. Volunteers after winter annuals.

Pests: Fall armyworm.

Caucasian Bluestem

Caucasian Bluestem
(Bothriocloa caucasica)

Origin: Former USSR
Description: Perennial bunchgrass, leafy with fine stems, deep-rooted and drought-tolerant. Grows 2 to 4 feet tall.
Primary Adaptation: Western part of Zones B, C, and D. Best on loam and clay soils.
Major Uses: Pasture, hay. Less palatable than big bluestem.
Establishment: Seed planted at 2 to 3 lb/A in May or June.
Fertilization: Responsive to nitrogen.
Seasonal Production: Late May-August.
Management: Stock heavily in early summer when quality is best. Nutritive quality declines with maturity.
Pests: No major pests.

Corn

Corn *(Zea mays)*

Origin: Southern Mexico
Description: Annual. Erect, thick-stemmed, leafy, 8 to 10 feet tall. High forage digestibility.
Primary Adaptation: Zones A, B, C, and D. Best adapted on loam soils with good water-holding capacity or irrigation.
Major Use: Silage, grain.
Establishment: Seed are drilled in wide rows at 12 to 20 lb/A in March-May.
Fertilization: High fertility. Very responsive to nitrogen and potassium. Needs lime on very acid soils.
Seasonal Production: Harvested July-October.
Management: Harvested for silage when kernels are in early dent stage.
Pests: Corn borer, corn rootworm, and corn smut.

Dallisgrass

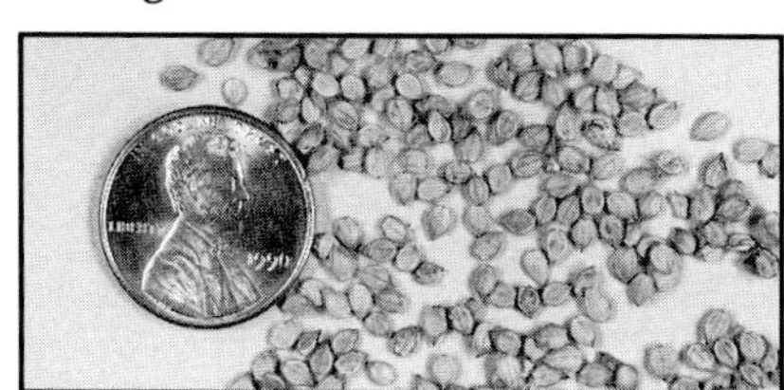

Dallisgrass *(Paspalum dilatatum)*

Origin: Argentina, Uruguay, Brazil.

Description: Perennial bunchgrass, short rhizomes. Grows 10 to 20 inches tall. Very leafy.

Primary Adaptation: Zone A, lower part of Zone B. Best adapted on clay and loam soils in areas of good summer moisture.

Major Uses: Pasture, but can be harvested for hay. Better nutritive quality than bermudagrass.

Establishment: Seed germination is low and establishment is slow. About 10 to 15 lb/A of pure live seed should be broadcast planted in March or April.

Fertilization: Moderately responsive to nitrogen.

Seasonal Production: April-October.

Management: Best grown with ladino or red clover. Seed heads can be clipped to eliminate ergot problem if it develops.

Pests: Ergot reduces seed set and can cause toxicity (usually mild) in cattle. Sugarcane borer.

Dallisgrass field scene.

Eastern Gamagrass

Eastern Gamagrass
(Tripsacum dactyloides)

Origin: Eastern Great Plains and eastern USA.
Description: Perennial bunchgrass, short rhizomes, 3 to 8 feet tall.
Primary Adaptation: Western part of Zones C and D. Loam to clay soils.
Major Uses: Pasture, hay. Good nutritive quality. Little is planted.
Establishment: Commercial seed production is a problem and establishment is difficult.
Fertilization: High fertility.
Seasonal Production: June-August.
Management: Rotational stocking.
Pests: Unknown.

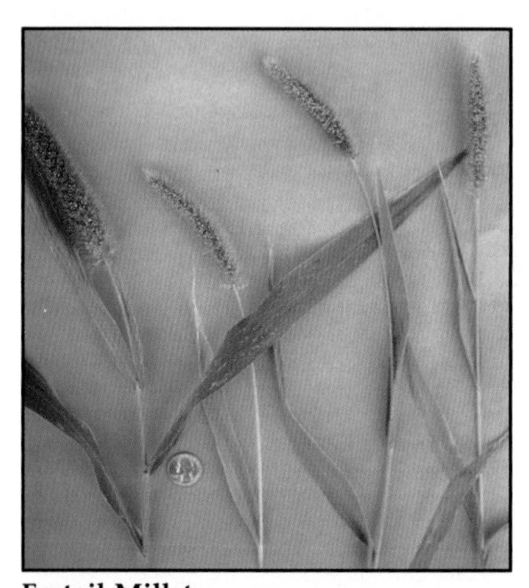

Foxtail Millet

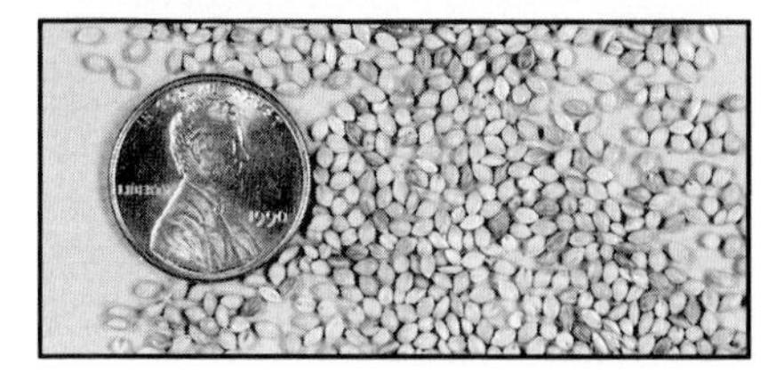

Foxtail Millet *(Setaria italica)*

Origin: Southern Asia
Description: Annual. Erect, 3 to 4 feet tall, leafy, fine-stemmed.
Primary Adaptation: Well-drained soils in Zones A, B, C, and D. Fairly drought-tolerant.
Major Uses: Hay. Once was widely grown, but infrequently at present.
Establishment: Seed can be drilled at 15 to 20 lb/A or broadcast at 20 to 30 lb/A in May-July.
Fertilization: Responds to nitrogen fertilization.
Seasonal Production: Most varieties are ready to be harvested in 60 to 70 days. Less productive than pearl millet or sorghum-sudan hybrids.
Management: Should be cut for hay near seedhead emergence. Not recommended for horse hay because of a toxin which can cause kidney and joint problems.
Pests: Unknown.

Indiangrass

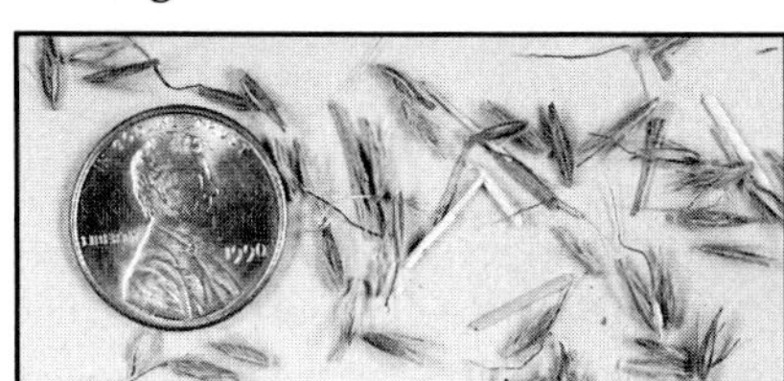

Indiangrass *(Sorghastrum nutans)*

Origin: Native to tall grass prairie of eastern Great Plains and eastern USA.

Description: Perennial bunchgrass. Spreads by rhizomes and seed. Deep rooted. Grows 3 to 6 feet tall. The yellow panicles are 6 to 12 inches long.

Primary Adaptation: Best in western part of Zones C and D. Little is planted in other parts of the South at present. Well-drained, fertile clay soils. Heat and drought-tolerant.

Major Uses: Pasture, but can be harvested for hay. Nutritive quality is generally better than most other warm season perennial grasses.

Establishment: Seedlings grow slowly and compete poorly with weeds. Seed should be planted at 6 to 10 lb/A pure live seed during April-May.

Fertilization: Nitrogen is the most important fertilizer element. Response to other nutrients is generally lower than for cool season grasses.

Seasonal Production: Late June-September.

Management: Rotational stocking is essential if grazed below 6 inches. Can be continuously stocked if stubble is maintained between 10 to 16 inches. Very light grazing after September 1.

Pests: Minor.

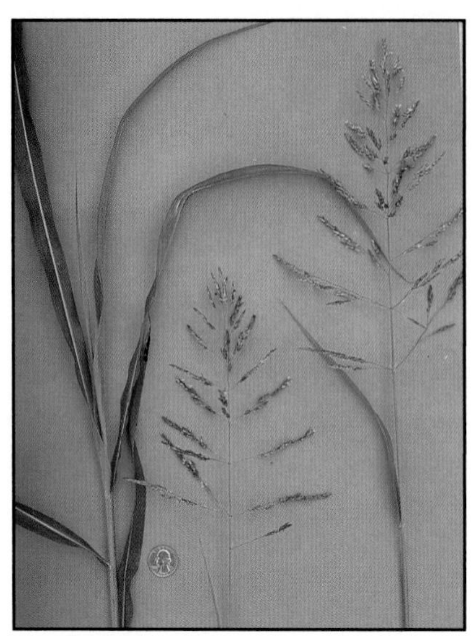

Johnsongrass

Johnsongrass ***(Sorghum halepense)***

Origin: Mediterranean region.

Description: Perennial. Erect, 3 to 6 feet tall. Spreads by rhizomes and seed. Drought-tolerant. Can be a serious pest in crops.

Primary Adaptation: Zones A, B, and C. Best on clay soils.

Major Uses: Best used for hay but can be grazed with good management.

Establishment: Seed are planted in April at 20 to 30 lb/A broadcast or 10 to 15 lb/A drilled.

Fertilization: Responds well to nitrogen.

Seasonal Production: May-September.

Management: Hay should be harvested at heading. If used for pasture, rotational stocking is needed to maintain stands. Close continuous grazing will reduce vigor and stands. Prussic acid potential (see Chapter 21).

Pests: No major pests.

Johnsongrass field scene.

Pearl Millet

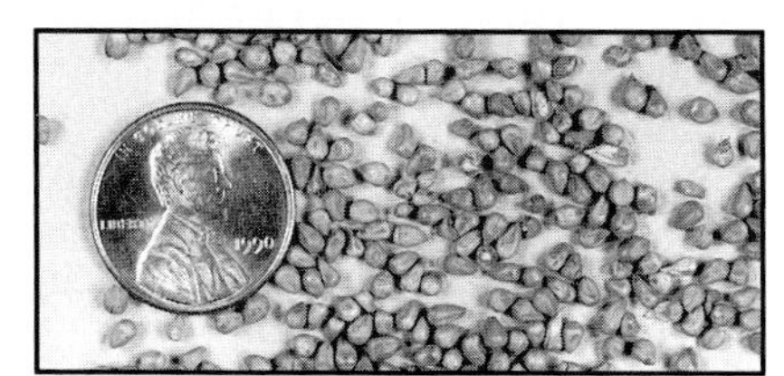

Pearl Millet ***(Pennisetum glaucum)***

Origin: North Central Africa.

Description: Annual. Erect, 3 to 8 feet tall, leafy.

Primary Adaptation: Zones A, B, and C. Best adapted on sandy soils. Does not do well on calcareous soils. Tolerant of drought and soil acidity.

Major Uses: Pasture, silage. Difficult to make hay because of thick stems. High nutritive quality if harvested at immature stage. Nitrate accumulation can cause toxicity under some circumstances (see Chapter 21).

Establishment: Seed are drilled at 12 to 15 lb/A or broadcast at 25 to 30 lb/A in April-June.

Fertilization: Much more tolerant of soil acidity than sorghum. Responsive to nitrogen.

Seasonal Production: Very productive over a short season, generally from June-August.

Management: Requires high stocking rate, preferably with rotational stocking. Stems may need to be mowed after grazing. Should be cut for hay when plants are 30 to 40 inches tall.

Pests: Fall armyworm.

Sorghum

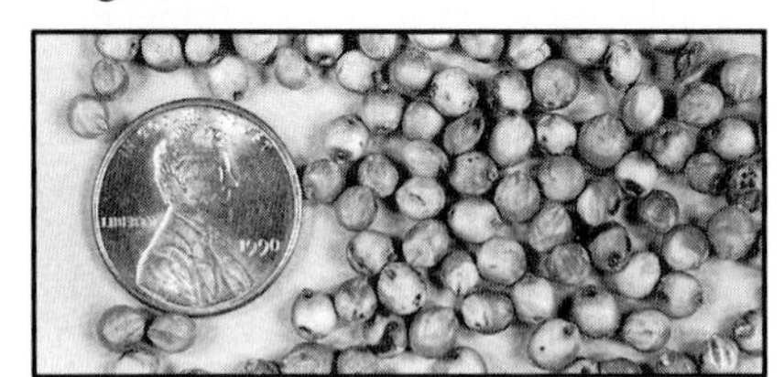

Sorghum *(Sorghum bicolor)*

Origin: Northeast Africa.

Description: Coarse-stemmed, erect annual, 4 to 15 feet tall. Grain types are short with large seed heads, forage types are tall with small seed heads.

Primary Adaptation: Zones A, B, C, and D. Very drought-tolerant. Not tolerant of highly acid soils.

Major Uses: Silage. Nutritive value is 85 to 90 percent of corn silage. Nitrate accumulation or prussic acid can cause toxicity under some circumstances (see Chapter 21).

Establishment: Seed are drilled in wide rows at 4 to 6 lb/A or broadcast at 15 to 20 lb/A in May-June.

Fertilization: Moderate fertility requirement.

Seasonal Production: Harvested July-August. Ratoon crop (second harvest) is possible in Zone A.

Management: Harvested for silage when seed are in early dough stage.

Pests: Leafspots, lesser cornstalk borer, sorghum midge.

Sorghum is a drought-tolerant crop that can be harvested for high-energy silage.

Sorghum-Sudan Hybrid

Sudangrass

Sorghum-Sudan Hybrids and Sudangrass *(Sorghum bicolor)*

Origin: Northeast Africa.
Description: Annual. Erect, 4 to 8 feet tall, leafy.
Primary Adaptation: Zones A, B, C, and D. Very drought-tolerant. Not tolerant of highly acid soils.
Major Uses: Pasture, hay, silage. High quality if harvested at immature stage. Difficult to make hay because of thick stems. Nitrate accumulation or prussic acid can cause toxicity under some circumstances (see Chapter 21).
Establishment: Seed are drilled at 20 to 25 lb/A or broadcast at 30 to 35 lb/A in May or June.
Fertilization: Very responsive to nitrogen. Needs lime on highly acid soils.
Seasonal Production: Quite productive over a short season. June-September.
Management: Requires high stocking rate, preferably grazed rotationally, to utilize rapid growth and maintain high quality. Thin-stemmed varieties recover more rapidly after cutting or grazing than thick-stemmed varieties. Stems may need to be mowed after grazing. Hay and silage should be cut when plants are 30 to 40 inches tall.
Pests: Fall armyworm.

Switchgrass

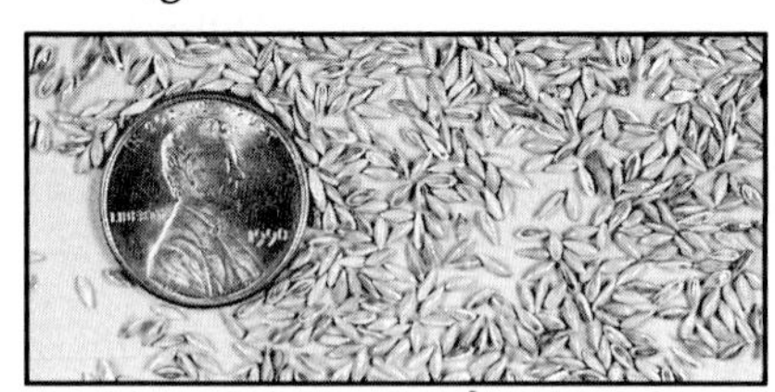

Switchgrass *(Panicum virgatum)*

Origin: Native to Great Plains and most of eastern USA.

Description: Perennial bunchgrass. Spreads by rhizomes and seed. Deep-rooted. Grows 3 to 7 feet tall.

Primary Adaptation: Zones B, C, and D. Drought-tolerant. Will tolerate poorly drained soils.

Major Uses: Pasture, hay. Develops stems several weeks earlier than other warm season grasses so may become stemmy and unpalatable early in summer. Improved varieties have higher yields and nutritive quality.

Establishment: Slow seedling establishment. Seed should be planted at 5 to 6 lb/A pure live seed in April-May.

Fertilization: Very responsive to nitrogen.

Seasonal Production: Late May-July.

Management: Should be stocked heavily and rotationally stocked with 4 to 6 weeks rest between grazings to maintain quality and stands.

Pests: Minor. ■

Chapter 6

Cool Season Grasses

COOL SEASON PERENNIAL GRASSES are the main pasture and hay species in regions north of Zone A (see **Figure 3.2**). They are generally of higher nutritive value and have a longer productive season than warm season perennial grasses. Cool season annual grasses are widely grown in Zones A and B where they provide high quality grazing when warm season grasses are dormant.

Kentucky Bluegrass

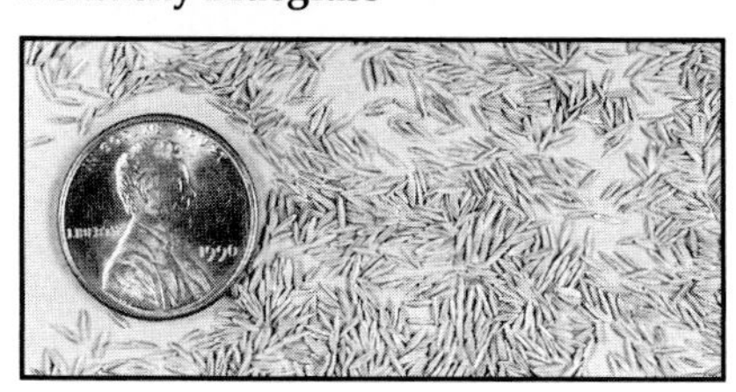

Kentucky Bluegrass *(Poa pratensis)*

Origin: Europe.

Description: Perennial. Rhizomes produce a dense sod. Grows 1 to 3 feet tall.

Primary Adaptation: Upper part of Zone C and Zone D. Intolerant of drought and high temperature.

Major Uses: Pasture, limited use for hay. High nutritive value.

Establishment: Seed are planted at 10 to 15 lb/A in August-September, or sometimes in February-March.

Fertilization: Kentucky bluegrass will survive under low fertility, but is not highly productive unless well fertilized or grown with a legume.

Seasonal Production: April-October, limited in July and August.

Management: Kentucky bluegrass tolerates close and frequent grazing better than many grasses. Grazing to a height of 1 to 2 inches favors productivity and maintains a dense sod. Bluegrass pasture is much more productive when clover is grown with it.

Pests: Grubs, bluegrass billbug, sod webworms, chinch bugs, frit fly, and the greenbug aphid can be serious pests and reduce production or even injure stands.

Orchardgrass

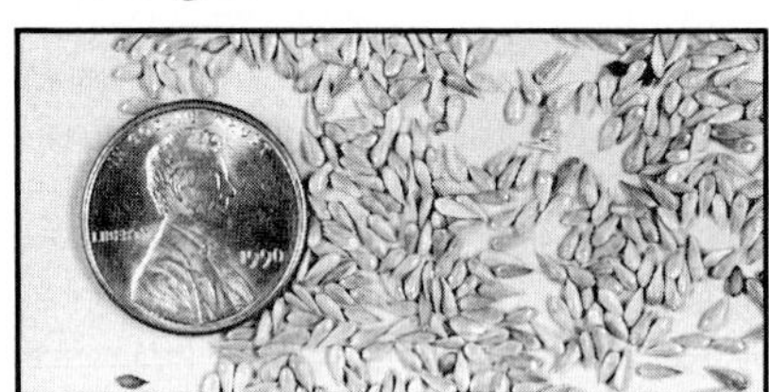

Orchardgrass *(Dactylis glomerata)*

Origin: Europe.

Description: Perennial bunchgrass. Grows 2 to 3 feet tall.

Primary Adaptation: Upper part of Zone B, and Zones C and D. Less tolerant of drought and poor drainage than tall fescue. Orchardgrass is more shade-tolerant than most grasses. In the lower South, stands do not generally persist more than 2 to 4 years.

Major Uses: Pasture, hay. Forage quality is high under good management.

Establishment: Seed should be planted in September at 15 to 20 lb/A, preferably with a legume. In upper part of Zone C and Zone D, plantings are sometimes made in early spring.

Fertilization: Requires higher fertility than tall fescue. Responds well to nitrogen.

Seasonal Production: March-June or July; production during September-November is much less than tall fescue.

Management: Orchardgrass requires better management than endophyte-infected tall fescue. Close continuous grazing will weaken stands. Moderate stocking is best. Orchardgrass is less competitive than tall fescue, so adapted legumes persist better in association with it. The first hay harvest of the season should be in the boot to early bloom stage, with subsequent harvests as growth permits.

Pests: Fall armyworm, rusts, and leafspot diseases. Nematodes are very serious on orchardgrass grown in sandy soils.

Prairiegrass

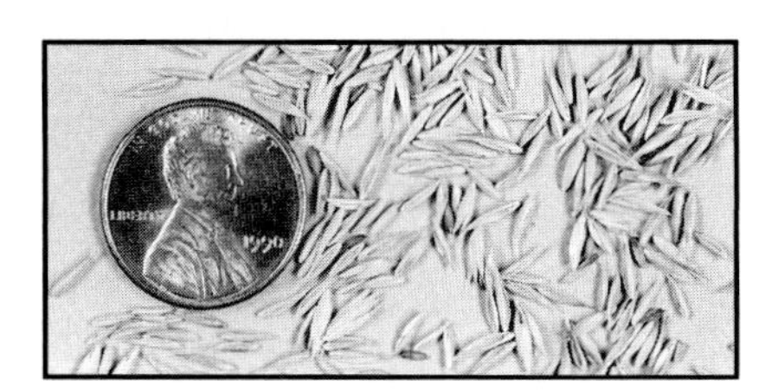

Prairiegrass*

(Bromus willdenowii)

Origin: Argentina.

Description: Short-lived perennial or natural reseeding annual bunchgrass. Grows 2 to 4 feet tall.

Primary Adaptation: Zones B, C, and D. Tolerant of drought. Not adapted to low fertility.

Major use: Pasture. High nutritive quality. Good winter production.

Establishment: Seed are planted in September at 25 to 30 lb/A, not deeper than ¼-inch. Seedling vigor and growth are exceptional.

Fertilization: High fertility requirement.

Seasonal production: November to May or June.

Management: Will not perenniate under close continuous grazing. Must be rotationally stocked to maintain stands.

Pests: Smut fungus on seed. Infected seed greatly reduces stand establishment and forage yield. Powdery mildew on forage.

*Also known as Matua bromegrass.

Reed Canarygrass

Rescuegrass

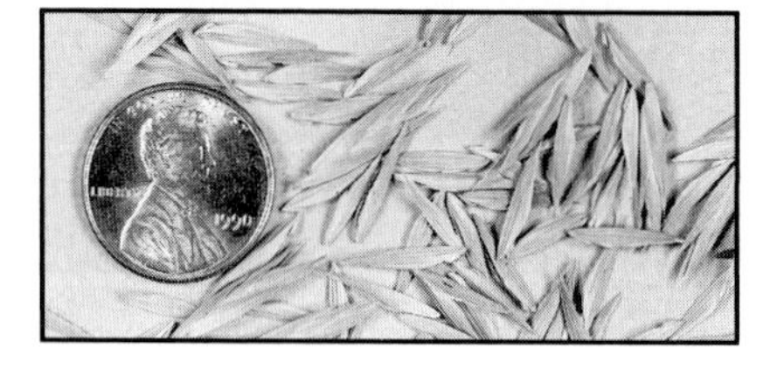

Reed Canarygrass
(Phalaris arundinacea)

Origin: Europe.
Description: Coarse, sod-forming perennial with short rhizomes, 2 to 6 feet tall. Deep-rooted.
Primary Adaptation: Upper part of Zone C and Zone D. Extremely tolerant of flooding and poorly drained soils. Drought-tolerant. Tolerates a soil pH range of 5 to 8.
Major Uses: Pasture, hay, silage.
Establishment: Seedling vigor is poor so establishment is slow. A seeding rate of 5 to 8 lb/A planted during April-May in Zone D or August-September in Zone C should give satisfactory stands.
Fertilization: Highly responsive to nitrogen.
Seasonal Production: April-September.
Management: Rotational stocking.
Pests: Tawny blotch, *Helminthosporium* leafspot, frit fly.

Rescuegrass
(Bromus catharticus)

Origin: Mediterranean area.
Description: Short-lived perennial bunchgrass that usually grows as an annual in the South. Grows 2 to 4 feet tall.
Primary Adaptation: Zones A, B, and C. Common rescuegrass is intolerant to extreme cold. Not adapted to low fertility.
Major Uses: Pasture. High nutritive quality.
Establishment: Seed are planted in September at 20 to 30 lb/A.
Fertilization: High fertility requirement.
Seasonal Production: November to May.
Management: Stocking rate should be high enough to utilize forage and reduce mildew problem.
Pests: Highly susceptible to mildew disease.

Annual Ryegrass

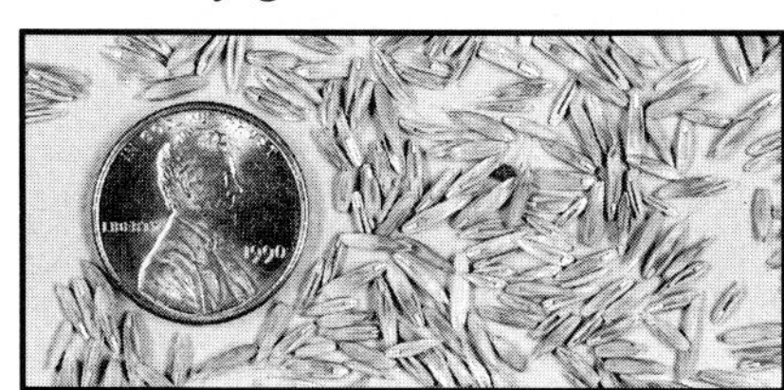

Ryegrass, Annual
(Lolium multiflorum)

Origin: Europe.

Description: Annual bunchgrass. Shiny, dark green smooth leaves. Grows 2 to 3 feet tall.

Primary Adaptation: Zones A and B. Tolerates wet, poorly drained soil. Less winter-hardy than tall fescue or orchardgrass. High moisture requirement.

Major Uses: Mainly pasture although sometimes used for hay or silage. High nutritive quality.

Establishment: Seeding rate is 10 to 15 lb/A in mixtures, or 20 to 30 lb/A alone. September or early October are generally the best months to plant, but November overseeding of warm season grasses can be done along the Gulf Coast. Natural reseeding is common.

Fertilization: Responsive to nitrogen. Tolerant of moderate soil acidity.

Seasonal Production: In high rainfall areas of the Gulf Coast, high production can be expected through the winter from November to May. Farther north, most of the production is concentrated from late February or March through May. Under favorable conditions, high forage production and excellent animal gains can be achieved.

Management: Ryegrass may be seeded alone in the Gulf Coast area but farther north it is usually seeded with a small grain (rye, wheat, or oats) and/or a clover. Ryegrass will tolerate close continuous grazing.

Pests: Armyworms. Rust disease attacks some varieties, especially within 100 miles of the Gulf of Mexico.

Rye

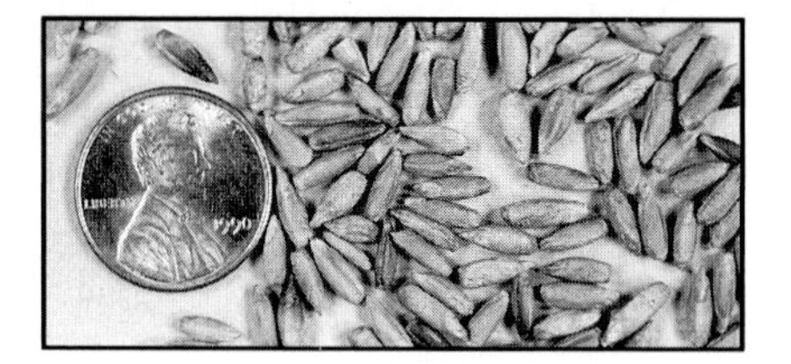

Oats

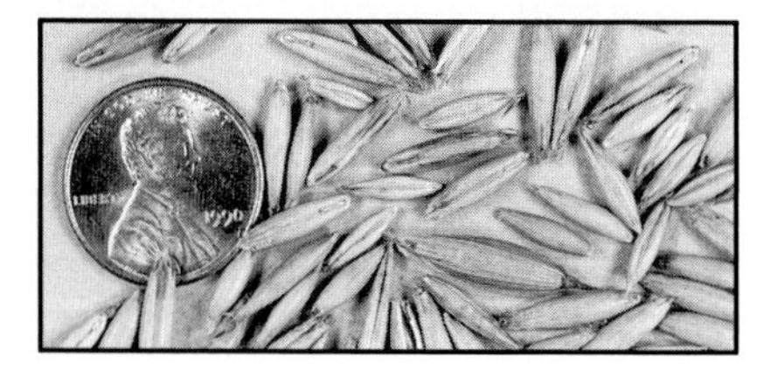

Small Grains
Rye *(Secale cereale)*
Oats *(Avena sativa)*
Wheat *(Triticum aestivum)*
Barley *(Hordeum vulgare)*
Triticale *(Triticum secale)*

Origin: Iraq, Turkey, Europe.

Description: Annual bunchgrasses 2 to 4 feet tall.

Primary Adaptation: Rye and wheat in Zones A, B, C, and D, oats Zone A. Rye is more tolerant of soil acidity than wheat or oats. Although varieties differ, oats generally is more cold sensitive than other small grains and can be winterkilled some years.

Major Uses: Rye–pasture; Barley, Wheat, Oats–pasture hay, silage; Triticale–hay, silage.

Establishment: Seed are usually planted in September or October. In mixtures, 60 to 90 lb/A are recommended, but 90 to 120 lb/A if planted alone.

Fertilization: All the small grains are highly responsive to nitrogen and require adequate amounts of phosphorus and potassium.

Seasonal Production: Zone A–November to April; Zone B–November to December and February to April; Zones C and D–March to June.

Wheat

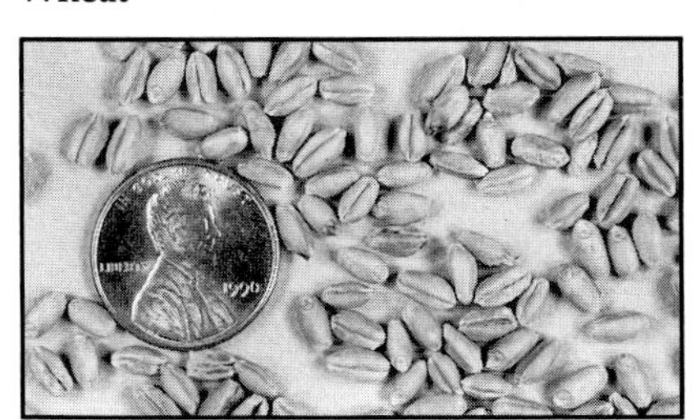

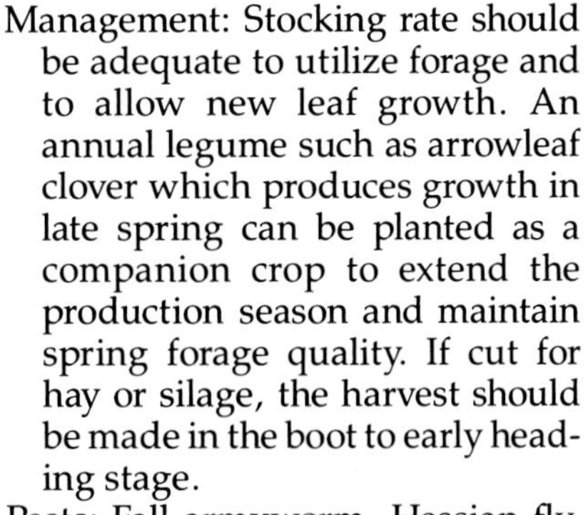

Management: Stocking rate should be adequate to utilize forage and to allow new leaf growth. An annual legume such as arrowleaf clover which produces growth in late spring can be planted as a companion crop to extend the production season and maintain spring forage quality. If cut for hay or silage, the harvest should be made in the boot to early heading stage.

Pests: Fall armyworm, Hessian fly.

VEGETATIVE IDENTIFICATION OF SMALL GRAINS

Identification of the vegetative growth of small grains can be difficult, but the leaf collar of most varieties of a given species exhibits unique characteristics. Triticale tends to be more variable than the other small grains.

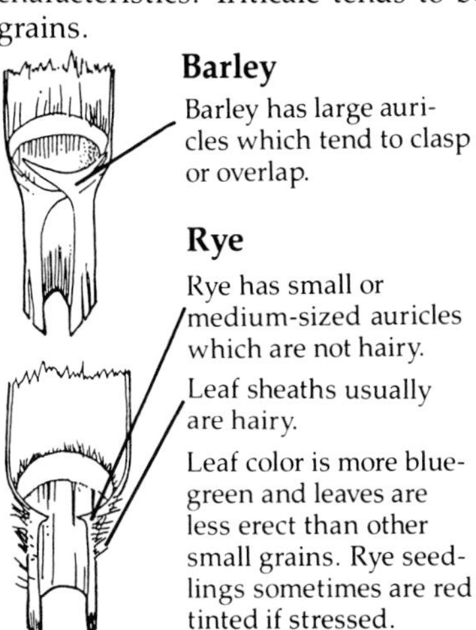

Barley

Barley has large auricles which tend to clasp or overlap.

Rye

Rye has small or medium-sized auricles which are not hairy.

Leaf sheaths usually are hairy.

Leaf color is more blue-green and leaves are less erect than other small grains. Rye seedlings sometimes are red tinted if stressed.

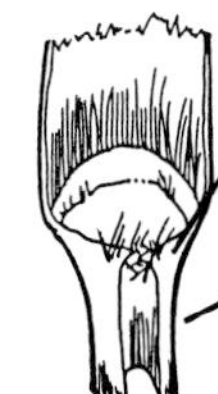

Wheat

Wheat has small or medium-sized auricles which are hairy.

The leaf sheath is not hairy.

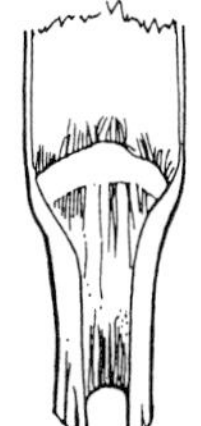

Oats

Oats have no auricles. Leaves of most varieties are wider than those of other small grains.

Figure 6.1. Vegetative identification of small grains.

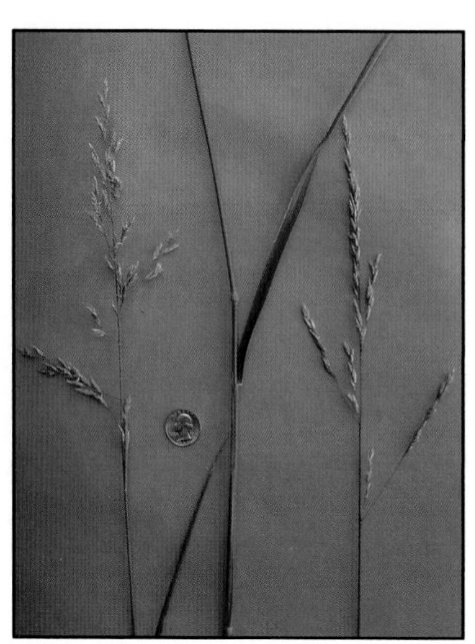

Tall Fescue

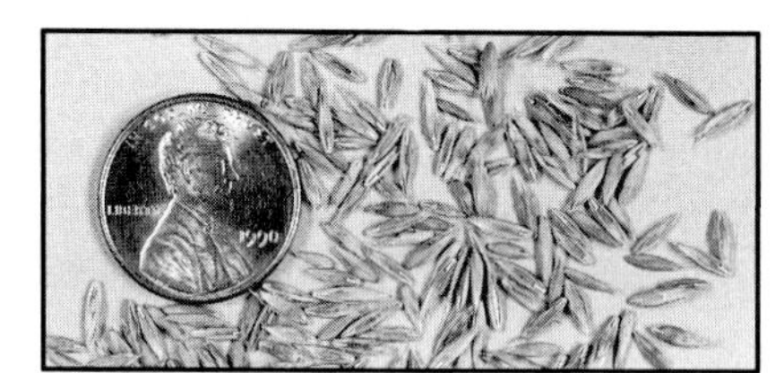

Tall Fescue *(Festuca arundinacea)*

Origin: Europe.

Description: Perennial long-lived bunchgrass with short rhizomes. Shiny, dark green, ribbed leaves. Deep rooted. Grows 2 to 4 feet tall.

Primary Adaptation: Zones B, C, and D. Best adapted in clay or loam soils. Tolerant of soil acidity and poor drainage, and relatively tolerant of drought.

Major Uses: Pasture, hay, erosion control. High forage quality if free of fungal endophyte (see Chapter 23).

Establishment: Seed are drilled at 15 to 20 lb/A or broadcast at 20 to 25 lb/A in September or October. In Zones C and D, planting can be done in early spring.

Fertilization: Tolerant of low fertility and acid soils but responds well to fertilization.

Seasonal Production: September-December, March-June or July.

Management: Tall fescue, if endophyte-infected, will tolerate grazing abuse better than most grasses. Endophyte-free tall fescue should not be grazed closer than 3 inches and should especially not be overgrazed during summer. Ladino or red clovers or alfalfa can be grown with tall fescue. The first harvest of hay should be cut in the late boot stage for high quality. Subsequent harvests can be made as growth permits.

Pests: The fungal endophyte *Acremonium coenophialum* can drastically reduce the performance of animals consuming tall fescue pasture or hay (see Chapter 23). Only endophyte-free seed should be planted if high animal performance is desired. Nematodes are a serious problem of tall fescue on sandy soils.

Timothy

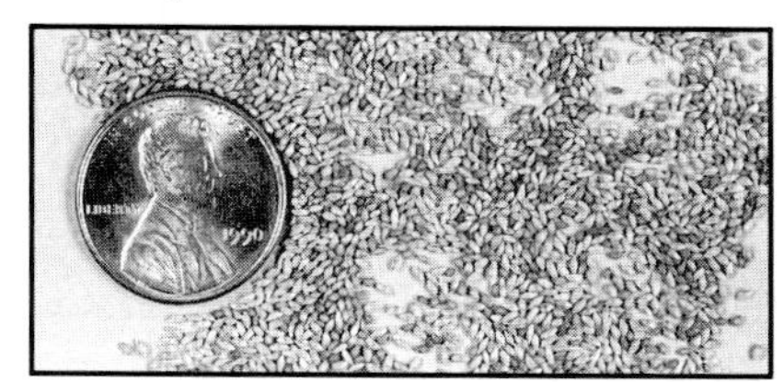

Timothy *(Phleum pratense)*

Origin: Northern Europe.

Description: Perennial bunchgrass. Grows 2 to 4 feet tall.

Primary Adaptation: Upper part of Zone C and Zone D. Cool, humid conditions without drought. When planted in warmer zones, timothy stands weaken and disappear quickly.

Major Use: Primarily as a hay plant but also used for pasture. It is popular as a hay crop for horses although other grasses are equally satisfactory.

Establishment: Seed are planted at 6 to 8 lb/A with a clover, alfalfa, or trefoil in August or September, or sometimes in early spring.

Fertilization: Tolerant of low fertility but responds well to fertilization.

Seasonal Production: April-October with low production in August-October.

Management: Hay should be cut at the boot to early bloom stage to obtain best quality. Late-cut timothy hay is of low quality.

Pests: No serious pest problems when grown in area of adaptation. ■

Chapter 7

Warm Season Legumes

THERE IS A GREAT NEED in Southern livestock production for persistent perennial legumes, especially warm season perennial legumes, that provide good quality forage and fix nitrogen over an extended portion of the year. Two of the most useful perennial legumes in the South, alfalfa (a cool season species which grows throughout the summer), and improved sericea lespedeza, have quite different adaptation requirements and uses. Both legumes are grown on a relatively small but increasing acreage as producers recognize their valuable characteristics. It is unlikely that the South will achieve its full potential in livestock production without greatly expanded acreage of these two valuable perennial legumes. Adaptation zones are provided in **Figure 3.2.**

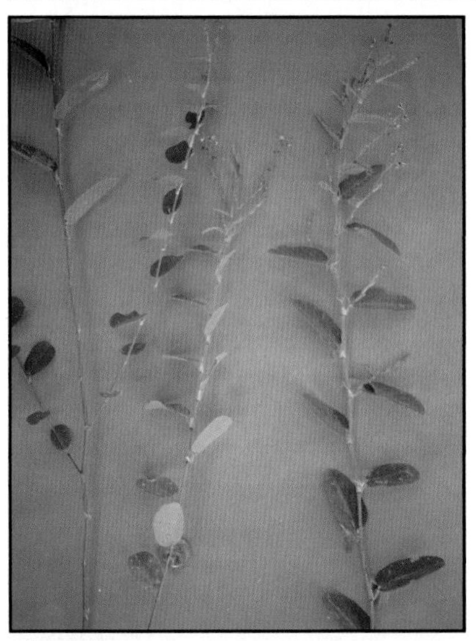

Alyceclover

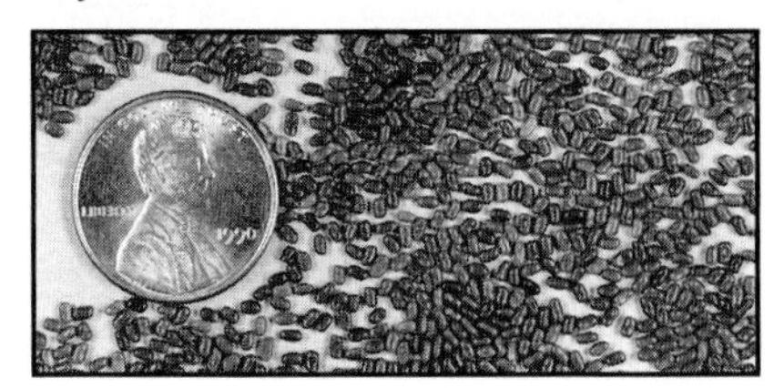

Alyceclover
(Alysicarpus vaginalis)

Origin: Tropical areas of the Orient.

Description: Annual. Erect, thin-stemmed, rounded leaves, pink flowers, 12 to 24 inches tall. Not a true clover.

Primary Adaptation: Zone A, grows best in Gulf Coast area with high summer rainfall.

Major Uses: Pasture, hay. High nutritive quality. Maintains quality well in late summer.

Establishment: Seed are planted at 15 to 20 lb/A in May-June. Establishment is slow and weed competition may be a problem.

Fertilization: Tolerant of soil acidity but responds to phosphorus fertilization.

Seasonal Production: July-September.

Management: Best adapted to well-drained, sandy soils. Grazing should begin at 12 to 15 inches. Hay should be cut at 18 to 24 inches. A second cutting of hay is possible under favorable growing conditions. Reseeding is not dependable.

Pests: Susceptible to nematodes. Not competitive with weeds in seedling stage.

Striate Lespedeza

Korean Lespedeza

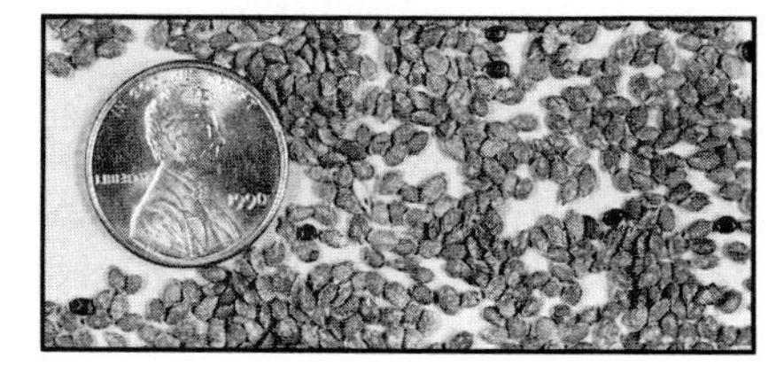

Annual Lespedeza
Striate *(Kummerowia striata)* and Korean *(Kummerowia stipulacea)*

Origin: Eastern China, Korea, Japan.

Description: Annuals. Both species are fine-stemmed, leafy, and have shallow taproots. Leaflets of striate are narrower than Korean. Seed of striate are in leaf axils while seed of Korean are borne in clusters at tips of branches. Reseed easily.

Primary Adaptation: Korean in Zones B, C, and D, striate in Zones A, B, and C.

Major Uses: Pasture, hay, erosion control. High nutritive quality, furnishing excellent quality pasture in late summer. Relatively low yield, especially in southern zones.

Establishment: Seed are planted at 25 to 35 lb/A in February-March.

Fertilization: Tolerant of acidity and low soil phosphorus. Under high fertility, annual lespedezas are often crowded out by more vigorous and higher yielding grasses and legumes.

Seasonal Production: July-September.

Management: High fertilization of a grass/lespedeza mixture reduces potential growth of lespedeza. Light grazing will allow some seed production for natural reseeding. Hay should be cut at early bloom stage.

Pests: Bacterial wilt, tar spot, powdery mildew, and southern blight can reduce yield and stands (striate is more resistant to these diseases than Korean). Insect damage is slight.

Cowpea

Cowpea *(Vigna unguiculata)*

Origin: Ethiopia.

Description: Annual. Viney, weak stems, large leaves, and curved pods.

Primary Adaptation: Zones A, B, and C. Tolerant of drought, low fertility, and soil acidity. Needs good drainage.

Major Uses: Pasture, hay. Forage quality is good.

Establishment: Seed are planted in rows at 30 to 40 lb/A, or 100 to 120 lb/A broadcast in May to early June.

Fertilization: Tolerant of low fertility and soil acidity.

Seasonal Production: June-August.

Management: Similar to soybeans.

Pests: Grasshoppers, curculio weevil, Southern cornstalk borer, and armyworm can be serious insect pests. Rusts and leafspots.

Kudzu

Note: Kudzu does not usually produce mature seed. It is propagated vegetatively from crowns.

Kudzu *(Pueraria lobata)*

Origin: Japan, Korea, eastern China.

Description: Perennial. Hairy, large leaves on long, twining vines arising from heavy woody crown. Reddish purple flowers rarely make seed before frost in the South. Spreads rapidly. Deep-rooted and drought-tolerant.

Primary Adaptation: Zones A, B, and C. Rarely planted but volunteers in many parts of the South.

Major Uses: Erosion control, pasture, hay. Good nutritive quality. Can be a pest where not grazed or otherwise controlled.

Establishment: Vegetative from crowns.

Fertilization: Tolerant of soil acidity and low fertility.

Seasonal Production: May-October.

Management: Usually only one hay cutting can be made per season. If pastured, rotational stocking should be used to maintain crown and root food reserves. Continuous close grazing will weaken plants, reduce productivity, and eventually eliminate stands.

Pests: No serious pests.

Perennial Peanut

Note: Perennial peanut produces few seed. It is propagated vegetatively from rhizomes.

Perennial Peanut
(Arachis glabrata)

Origin: Brazil.

Description: Perennial, long-lived, leafy, 12 to 16 inches tall, spreading by rhizomes, yellow or orange flowers.

Primary Adaptation: Lower part of Zone A. Can survive temperatures down to 15°F, but a long cool season greatly reduces potential production. Grows best on well-drained sandy soils. Will not tolerate poor drainage.

Major Uses: Pasture, hay. High nutritive quality. Generally, 2 years are required for establishment of a productive stand.

Establishment: Rhizomes are planted at about 60 to 80 bu/A from December to early March. Perennial grasses can be interplanted after the peanuts are well established.

Fertilization: Low fertility requirement. Calcium application generally necessary.

Seasonal Production: April-October.

Management: Continuous stocking to maintain a height of at least 4 inches. Rotational stocking should allow at least a 3-week rest between grazing periods of 10 days or less. Two to three hay cuts can be obtained per year. No cutting should be made 5 to 6 weeks before killing frost to allow replacement of rhizome food reserves.

Pests: Minor.

Sericea Lespedeza

Sericea Lespedeza
(Lespedeza cuneata)

Origin: Eastern China, Korea, Japan.

Description: Perennial. Erect-growing, leafy, with fine stems in improved varieties. Grows 18 to 40 inches tall. Deep-rooted and drought-tolerant. Small flowers. Non-bloating legume.

Primary Adaptation: Zones A, B, C, and D. Best on clay or loam soils. Very tolerant of acid soils high in aluminum. Not adapted on calcareous or wet soils. Grows well on low-fertility soils where most other legumes do not thrive.

Major Uses: Hay, erosion control, pasture. Most varieties have high levels of tannin which reduce digestibility, but low-tannin varieties are available.

Establishment: Ideally, seed are planted at 20 to 30 lb/A with a preplant herbicide during late March-May. Establishment is often slow. Post-emergence herbicide treatment(s) may also be required.

Fertilization: Very tolerant of low fertility and acid soils; may respond to potassium on some soils.

Seasonal Production: April to September.

Management: Hay harvests which leave a 4-inch stubble height should be made when plants are 15 to 24 inches tall, obtaining 2 to 3 cuttings per year. High-tannin sericea forage has nutritive value similar to bermudagrass; low-tannin types have superior nutritive value. When cut for hay, the tannin level of forage drops sharply, improving palatability and digestibility. Vigor and yield of low tannin sericea tend to be somewhat less than for high-tannin types. Grazing of sericea should begin when plants are 8 to 10 inches tall, and they should not be grazed lower than 4 inches. Removal of forage by grazing or hay production between late August or early September and the first killing frost should be avoided, as this is a period when food reserves are building in the roots. Tall fescue or orchardgrass can be overseeded on sericea to provide a longer productive season than sericea alone. Winter annuals such as cereals, crimson clover, or vetch can also be drilled into sericea in October to provide late winter forage. Grasses overseeded on sericea should be fertilized with nitrogen.

Pests: Relatively free of insects and diseases. Dodder, a parasitic plant, can be a serious problem in hay or seed fields, so it is important to plant dodder-free seed and kill any volunteer dodder plants with a herbicide.

Soybean (Laredo variety)

Velvetbean

Soybean *(Glycine max)*

Origin: China

Description: Annual. Bushy, leafy plants.

Primary Adaptation: Zones A, B, and C. Tolerates drought when grown for forage. Well-drained soil.

Major Uses: Most varieties were developed for oil seed production but can be used for high quality short-season grazing or hay with yields of 2 to 3 tons/A. Hay should be harvested when 75 percent of pods are filled.

Establishment: Seed are planted at 60 to 100 lb/A in May.

Fertilization: Intolerant of soil acidity and low fertility.

Seasonal Production: July-August.

Management: One hay cut can be made. Hay is difficult to cure and requires a hay conditioner. If grazed, the season will be short as no regrowth can be expected.

Pests: Armyworm.

Velvetbean
(Stitzolobium deeringianum)

Origin: India

Description: Viney annual, extending 20 to 40 feet. Hairs on 2 to 6 inch pods can irritate human skin.

Primary Adaptation: Zone A. Sandy soils.

Major Uses: Pasture, green manure.

Establishment: Seed are planted at 30 lb/A after all danger of frost is past.

Fertilization: Tolerant of soil acidity and low fertility.

Seasonal Production: May-August.

Management: Moderate grazing to allow regrowth.

Pests: Velvetbean caterpillar. ■

Chapter 8

Cool Season Legumes

COOL SEASON LEGUMES make most of their growth in the winter and spring when temperatures and rainfall are generally favorable. Perennial legumes, which usually do not persist more than 2 or 3 years in the South, are generally restricted to Zones B, C, and D (see **Figure 3.2**) except in favorable soils with good water-holding capacity or a high water table in Zone A.

Annual legumes are most commonly grown in Zone A and the lower part of Zone B. Annuals have the disadvantage of needing to be reestablished each autumn. The seedling stage is the most hazardous period of development when plants are subject to damage from insects, disease, and drought. Thus, dependability is increased by using perennial legumes whenever possible. White clover and red clover are the perennial legumes most commonly grown in association with perennial grasses such as tall fescue, orchardgrass, or dallisgrass. Annual legumes are usually grown with annual grasses (a small grain and/or annual ryegrass) and may be planted on a prepared seedbed in early autumn or overseeded in bermudagrass or bahiagrass during autumn.

Alfalfa

Alfalfa *(Medicago sativa)*

Origin: Iran and central Asia.

Description: Perennial. Erect-growing with many leafy stems arising from large crowns at the soil surface. Grows 24 to 36 inches tall. Compound leaves with three leaflets. Flowers of the varieties grown in the South are normally some shade of purple. Drought-tolerant, long taproot.

Primary Adaptation: Zones B, C, and D, with a few varieties adapted in Zone A. Requires well-drained soils. Alfalfa planted on land having a very acid subsoil with large amounts of toxic aluminum will have shallow root development and lower yield.

Major Uses: Hay and haylage, but has potential for expanded pasture use. Good alfalfa hay has high nutritive value and is in high demand, particularly for horses and dairy cattle.

Establishment: Seed. A seeding rate of 15 to 20 lb/A should be used. A cultipacker-seeder is the best planting equipment for prepared seedbeds. A firm seedbed is critically important. For sodseeding, a no-till drill is needed. In Zones A and B, September-October is the best time to plant. Autumn planting on prepared land should be done in August to early September in Zones C and D. Winter-spring seeding can be done in February to April in Zones C, D, and the upper part of Zone B.

Fertilization: Alfalfa is sensitive to soil acidity, so pH values 6.5 or above are required for high yields. Where subsoils are very acid and high in aluminum, it may be possible to offset the toxic subsoil syndrome and promote deeper root development through deep incorporation of lime or the application of gypsum at a rate of about 2 ton/A. Potassium, phosphorus, sulfur, and boron are the nutrients which usually need to be applied in order to obtain good alfalfa production. Alfalfa requires large amounts of potassium. Annual soil tests are critical in monitoring soil nutrient levels for this crop. Nitrogen fertilization is not needed, since alfalfa fixes large amounts of nitrogen if properly nodulated.

Seasonal Production: March-November in Zone A, April-October in Zones B, C, and D. Alfalfa has the longest productive season of any Southern-adapted legume.

Management: Successful alfalfa production requires a higher level of management than other forage crops. For hay production, 4 to 7 cuttings can be made each year depending on location. In the extreme southern portion of Zone A, it may even be possible to obtain 8 or 9 harvests in some years. Harvesting at the early bloom stage is the best compromise for obtaining acceptable forage and nutrient yields with good stand persistence. In Zones B, C, D stand life is often 3 to 5 years, but alfalfa may persist for up to 8 years if adequately fertilized and cut at the proper stage of growth. In Zone A, alfalfa stands on sandy Coastal Plain soils may remain productive for only 2 to 4 years. If a hay type alfalfa is to be used for pasture, it should be cross-fenced and rotationally stocked for 5 to 7 days, followed by a 20 to 35-day recovery period. Continuous stocking of hay-type varieties will deplete food reserves in the roots and cause rapid stand loss. Grazing-tolerant varieties can be continuously stocked, but production will be higher if rotational stocking is practiced. In the upper South, allow 4 weeks of growth in late summer to replenish root reserves. This may be grazed in late autumn.

Pests: Adapted varieties are fairly resistant to diseases. Sclerotinia crown and stem rot can be a problem the establishment year on autumn-seeded alfalfa. The most important insect problem is the alfalfa weevil, which can be controlled by spraying with an insecticide in the spring and/or early cutting. Leafhoppers may be a problem in summer but can be controlled by spraying with an insecticide. A good stand of well-fertilized alfalfa is strongly competitive against weeds, but weeds can be a problem in thin stands. Nematodes are a serious problem on non-resistant varieties in the sandy Coastal Plain of Zone A.

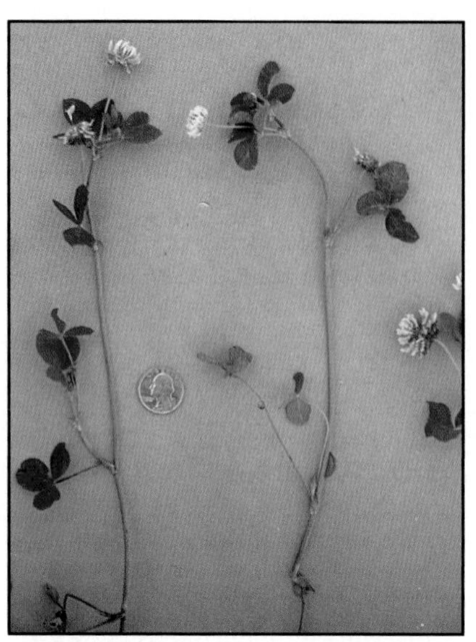

Alsike Clover

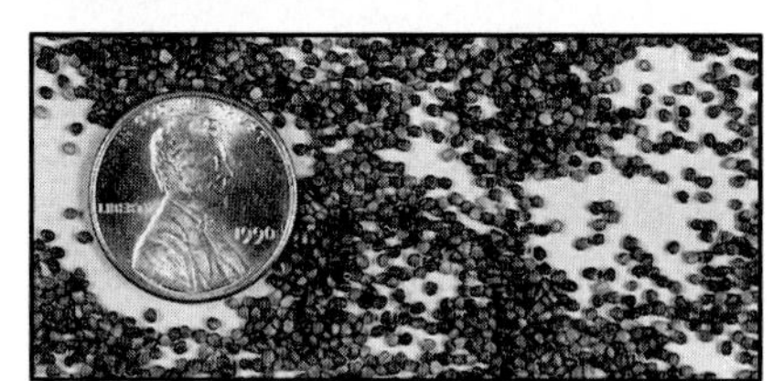

Alsike Clover ***(Trifolium hybridum)***

Origin: Northern Europe.

Description: Semi-erect, short-lived perennial, 1 to 3 feet tall, pink and white flowers.

Primary Adaptation: Cool climates and wet soils in Zones C and D. Tolerates more flooding and soil acidity than most clovers.

Major Use: Pasture, hay.

Establishment: Seed are broadcast at 4 to 6 lb/A.

Fertilization: Responds to phosphorus and potassium.

Seasonal Production: May-September.

Management: Stock heavy enough to prevent shading by grasses.

Pests: Powdery mildew, anthracnose, and several virus diseases.

Arrowleaf Clover

Arrowleaf Clover

(Trifolium vesiculosum)

Origin: Mediterranean region.

Description: Late season winter annual. Long, branching, hollow stems 2 to 4 feet long. The non-hairy arrow-shaped leaves generally have a large white "V" mark. The predominately white, but pink and purple-tinged flower heads are large, often 3 inches long. Flowering and seed production occur over a long period in late spring and summer. The rough brown seed are about twice the size of ladino clover seed and over 70 percent have very hard seed coats, requiring scarification for satisfactory germination.

Primary Adaptation: Zone A and lower part of B. Not adapted on calcareous soils or wet areas.

Major Uses: Pasture, hay. Forage quality is high with digestibility generally superior to crimson clover at all stages of maturity. Bloat is rarely a problem.

Establishment: Scarified seed are planted broadcast at 5 to 10 lb/A in September to early November. Requires a special seed inoculant. Seed will germinate at lower temperatures than crimson clover. Reseeds easily.

Fertilization: Arrowleaf is not very tolerant of soil acidity and low fertility. Optimum pH range is 5.8 to 6.5.

Seasonal Production: March-early July in Zone B, February-early June in Zone A.

Management: Arrowleaf will continue to develop new leaves and remain productive longer in the spring when grazed to a height of 2 to 4 inches than where large amounts of forage accumulate. Where hay is desired, clover should be grazed until early to mid-April, then harvested at early to mid-bloom in May. No regrowth can be expected after cutting hay.

Pests: Crown and stem rot may occur during warm, wet winter weather, especially on loam and clay soils. Grazing to remove surplus growth and to permit light and air movement in the sward will reduce disease losses. Nematodes and virus diseases can be problems as with most clovers. Arrowleaf clover leaves turn a distinctive purplish-red color in response to stress, whether caused by disease, nutritional imbalances, pest damage, or climatic conditions.

Ball Clover

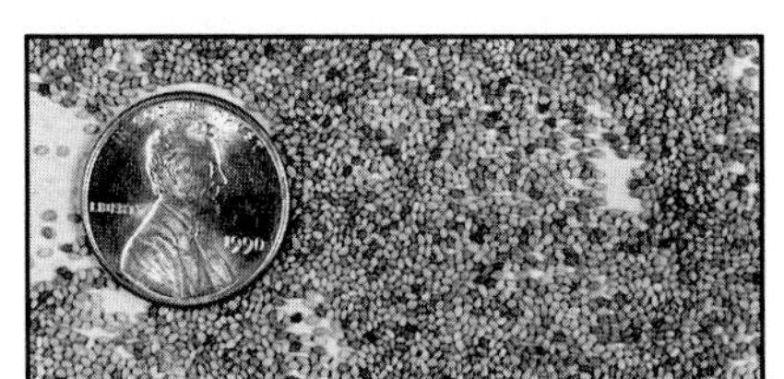

Ball Clover *(Trifolium nigrescens)*

Origin: Mediterranean region.

Description: Winter annual, prostrate to partially erect stems that reach a length of up to 3 feet tall. Fragrant white flowers in small heads. Seed are smaller than ladino clover. Hard seed and prolific seed production can result in good natural reseeding.

Primary Adaptation: Zone A and lower part of B. Loam to clay soils. Tolerates poor drainage.

Major Uses: Pasture, honey production. Bloat can be a problem.

Establishment: Seed are broadcast at 2 to 3 lb/A in September-October.

Fertilization: Similar to white clover. More tolerant of soil acidity than crimson clover.

Seasonal Production: Short season during late March-April.

Management: Tolerates heavy grazing and will produce seedheads close to the ground.

Pests: Clover head weevil may reduce seed production.

Berseem Clover

Berseem Clover
(Trifolium alexandrinum)

Origin: Mediterranean region.

Description: Winter annual, resembles alfalfa, hollow stems grow erect to a height of 2 feet or more. White flowers form small heads.

Primary Adaptation: Most berseem clover varieties are not winter-hardy in the southeastern USA. A winter-hardy variety can be grown in Zone A and the lower part of Zone B. Tolerant of alkaline soil. More tolerant of wet soil than most annuals. Particularly well-suited for use in non-acid Black Belt soils and in high rainfall areas near the Gulf Coast.

Major Uses: Pasture, hay. High quality, does not cause bloat.

Establishment: Seed are planted broadcast at 20 lb/A or drilled at 10 to 15 lb/A in September.

Fertilization: Best on loam soils of pH 6 or above. Requires high fertility including boron.

Seasonal Production: November-December and March-June.

Management: Grazing should begin when 10 inches tall and the stubble maintained at 3 to 4 inches to encourage new leaf production. Rotational stocking is most successful. The winter-hardy berseem variety produces hard seed and will often reseed.

Pests: Crown rot can be a problem if excess forage is present during freezing.

Birdsfoot Trefoil

Birdsfoot Trefoil
(Lotus corniculatus)

Origin: Mediterranean region.

Description: Deep-rooted, short-lived perennial, having finer stems and more leaves than alfalfa. Grows 12 to 30 inches tall, depending on whether it is a prostrate or erect variety. Flowers are bright yellow and the brown to purple seed pods radiate from the stem branch, resembling a bird's foot. Non-bloating legume. Reseeds under proper management.

Primary Adaptation: Upper part of Zone B, and Zones C and D. Well-drained soils. Tolerant of drought, moderate soil acidity. Grows well in association with Kentucky bluegrass, orchardgrass, and tall fescue. Will not compete with bermudagrass or bahiagrass.

Major Uses: Pasture, erosion control. High quality forage.

Establishment: Seed are planted at 4 to 6 lb/A with cool season perennial grasses using a cultipacker-seeder in late August-September.

Fertilization: Lime if soil pH is below 5.5. Responds well to phosphorus and potassium.

Seasonal Production: April-early October.

Management: Practical grazing management should leave 3 to 4 inches of leaf tissue to maintain root carbohydrates and assure vigorous regrowth. Since natural reseeding is essential for maintaining stands and productivity, plants should be allowed to produce some seed each year.

Pests: Crown and root rots are serious diseases of birdsfoot trefoil that reduce stands, requiring natural reseeding to maintain stands. Nematodes are a major problem on sandy soils.

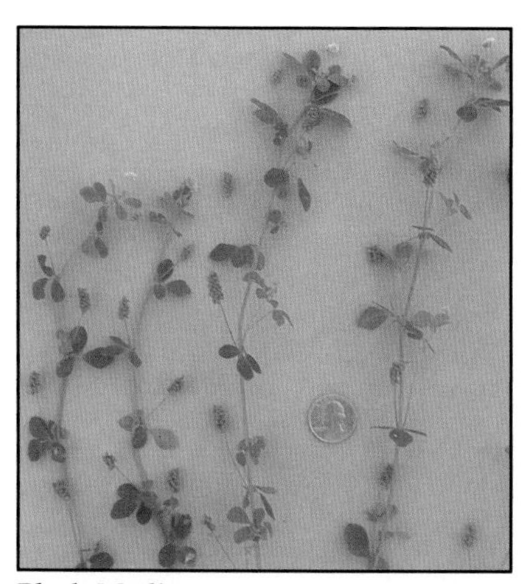

Black Medic

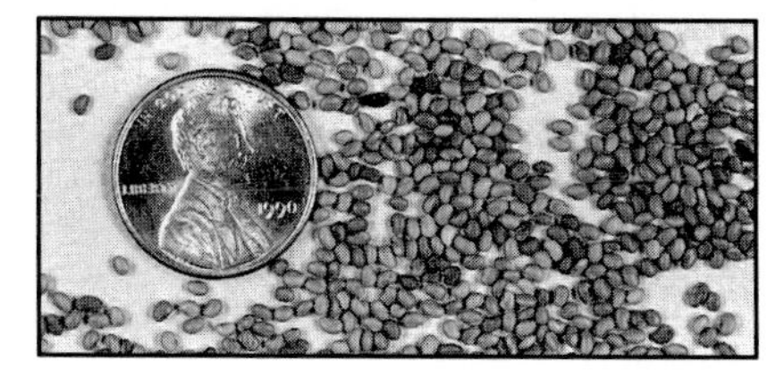

Black Medic *(Medicago lupulina)*

Origin: Mediterranean region.

Description: Winter annual with prostrate stems, usually not over 6 to 8 inches tall. Slightly hairy leaves. Small yellow flowers forming a bur.

Primary Adaptation: Zones A, B, C, and D.

Major Uses: Pasture.

Establishment: Volunteers in many areas. Seed are broadcast at 10 to 12 lb/A in September-October. Seed not readily available commercially.

Fertilization: Grows on moderately acid soil but is most productive on calcareous soils.

Seasonal Production: Low yields, occurring mainly in April and early May.

Management: Should be quickly utilized due to short productive season.

Pests: Unknown.

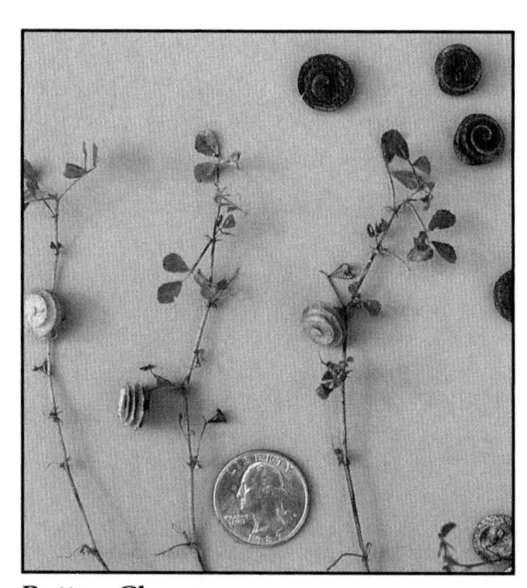

Button Clover

Button Clover *(Medicago orbicularis)*

Origin: Mediterranean region.

Description: Winter annual with leafy decumbent stems 2 to 3 feet long. Non-hairy leaves. Orange-yellow flowers that develop with flat, coiled seed pods that lack the spines of other burclovers.

Primary Adaptation: Zones A, B, and C on loam or clay soils. Best growth on limestone soils.

Major Uses: Pasture, hay.

Establishment: Seed are broadcast at 10 lb/A in September-October. Seed not readily available commercially.

Fertilization: Lime needed if soil is below pH 6.

Seasonal Production: March-May.

Management: If cut for hay, only one harvest can be obtained.

Pests: Unknown.

Caleypea, Roughpea, or Singletary Pea

Caleypea, Roughpea, or Singletary Pea
(Lathyrus hirsutus)

Origin: Mediterranean region.

Description: Winter annual with thin-winged spreading stems up to 3 feet long. Leaves non-hairy. Small red and blue flowers that form small seedpods. Hard seed coats can result in good reseeding.

Primary Adaptation: Zone A or lower part of Zone B on loam and clay soils. Grows on soils too wet for most clovers; also does well on both acid and calcareous soils.

Major Uses: Pasture. Seed are mildly toxic to livestock.

Establishment: Scarified seed are broadcast at 50 to 55 lb/A in September-October.

Fertilization: Phosphorus and potassium are required.

Seasonal Production: March-May.

Management: Grazing should be discontinued in late spring when seedpods are forming to avoid toxicity and to permit seed production for natural reseeding.

Pests: Aphids.

Crimson Clover

Crimson clover field scene.

Crimson Clover
(Trifolium incarnatum)

Origin: Mediterranean region.

Description: Winter annual. Plants have dark green leaves densely covered with hairs, and grow to a height of 1 to 3 feet. Brilliant crimson flowers, long heads, maturing from bottom to top. Yellow rounded seed about 2.5 times the size of arrowleaf clover seed. Combine-harvested seed do not need scarification.

Primary Adaptation: Mainly Zones A and B, limited use in Zone C. Does not tolerate calcareous or poorly drained soils.

Major Uses: Pasture, hay, green manure crop, roadside beautification. Will produce more forage at low temperatures than other clovers.

Establishment: Seed are broadcast at 20 to 30 lb/A in late August-October.

Fertilization: Fairly tolerant of soil acidity.

Seasonal Production: Late November-December, February-early April in Zone A; November, March-April in Zone B.

Management: Can be grazed throughout winter but if hay is desired, cattle must be removed by mid-March.

Pests: Clover head weevils often cause heavy seed losses, resulting in poor natural reseeding. Cool, wet weather in winter often results in crown and stem rot, especially where there is a thick layer of leaves. Moderate stocking in winter reduces the problem.

Lappa Clover

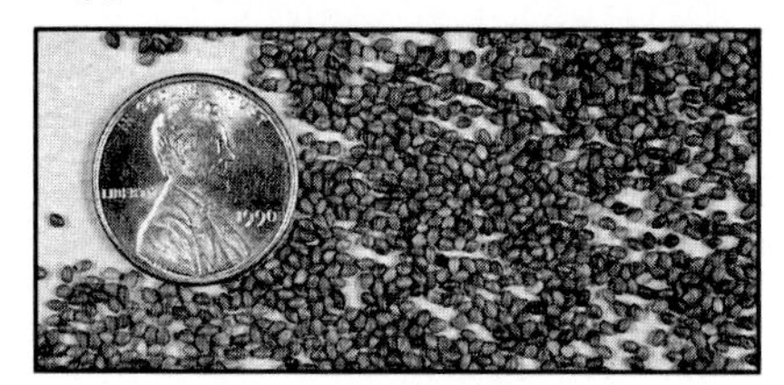

Lappa Clover
(Trifolium lappaceum)

Origin: Mediterranean region.

Description: Dense growth habit, growing to 2 feet. Leaves and stems very hairy. Pink-white flowers in small heads.

Primary Adaptation: Zone A and lower part of Zone B. Loam to clay soils. Tolerates poor drainage.

Major Uses: Pasture.

Establishment: Seed are broadcast at 10 to 15 lb/A in September-October. Seed not readily available commercially.

Fertilization: Grows best on calcareous soils.

Seasonal Production: April and May.

Management: Should be quickly utilized due to short productive period.

Pests: Unknown.

Large Hop Clover

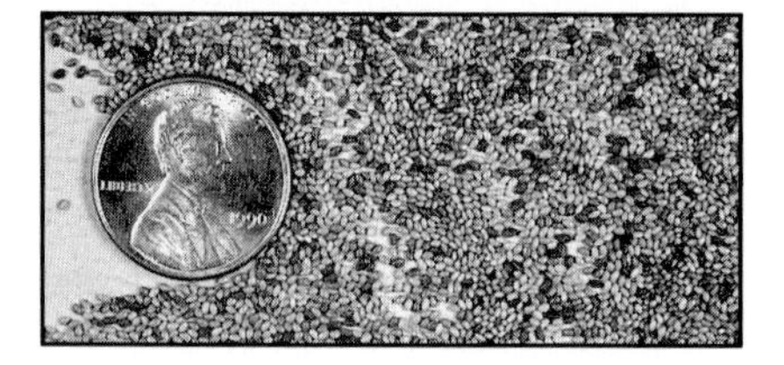

Large Hop Clover
(Trifolium campestre) and
Small Hop Clover
(Trifolium dubium)

Origin: Mediterranean region.

Description: Low-growing winter annuals. Yellow flowers in small heads.

Primary Adaptation: Zones A, B, C, and D.

Major uses: Pasture.

Establishment: Volunteers throughout the South. Seed are broadcast at 4 to 8 lb/A in September. Seed not readily available commercially.

Fertilizer: Tolerant of acid soils with low fertility.

Seasonal Production: Low yield over a short season in April and May.

Management: Should be quickly utilized due to short productive period.

Pests: Unknown.

Persian Clover

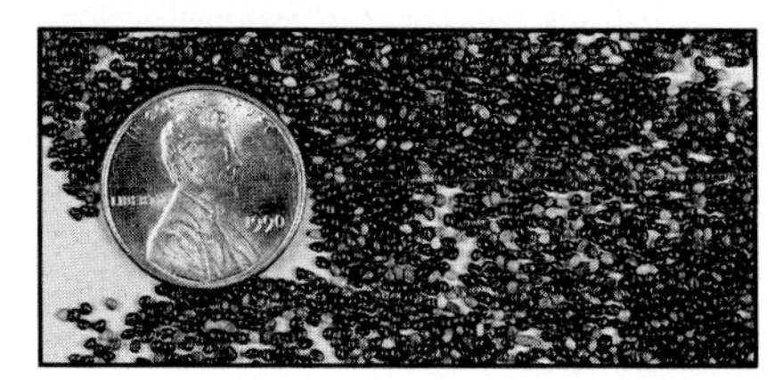

Persian Clover
(Trifolium resupinatum)

Origin: Mediterranean region.

Description: Winter annual, stems may reach a height of 2 to 3 feet. Small pink flowers. Seed are olive green to purple. Hard seed coats favor good reseeding.

Primary Adaptation: Zone A, lower part of Zone B. Loam to clay soils. Tolerates poor drainage.

Major Uses: Pasture, hay, green manure. Bloat can be a problem.

Establishment: Seed are broadcast at 3 to 5 lb/A in September-October.

Fertilization: Requires soil pH 6 or above.

Seasonal Production: Very productive in March-April.

Management: Close grazing.

Pests: No major pests.

Rose Clover

Rose Clover *(Trifolium hirtum)*

Origin: Mediterranean area of Europe and north Africa.

Description: Winter annual with spreading branches, hairy leaves, 18 inches tall, rose-colored flowers.

Primary Adaptation: Zone A and lower part of Zone B. Tolerant of drought and low soil fertility.

Major Uses: Pasture. High quality.

Establishment: Seeded broadcast at 15 to 20 lb/A in September-October. Good natural reseeding. Many seed are quite hard and can volunteer over many years.

Fertilization: Responds to phosphorus and potassium where these elements are deficient. Best adapted to soil pH 6 to 7.

Seasonal Production: California varieties are low yielders and mature in April in the South. An improved variety is available.

Management: Can be grazed continuously until maturity but livestock must be removed by mid-April if natural reseeding is desired.

Pests: Minor.

Red Clover

Red Clover ***(Trifolium pratense)***

Origin: Southeastern Europe and Turkey.

Description: Short-lived perennial, usually 2 years in the South, but may survive only as an annual in Zone A. Erect-growing, leafy plants 2 to 3 feet tall. Leaves hairy and marked with a white "V". Flowers are clustered into large pinkish-violet heads.

Primary Adaptation: Best in Zones C and D, but with good management it will perenniate in Zone B. Grows as a winter annual in Zone A. Fairly drought-tolerant. Tolerates more soil acidity and poorer soil drainage than alfalfa. Slightly less tolerant of extremely wet conditions than white clover. Due to a long growing season, red clover is the best yielding clover in areas where well adapted. Well suited for hay, silage, and grazing with tall fescue, orchardgrass, timothy, dallisgrass, or johnsongrass.

Major Uses: Hay, pasture.

Establishment: Seed are planted on prepared land at 6 to 8 lb/A in drill rows or 12 to 15 lb/A broadcast during September-October in Zones A, B, C, and D and February-March in Zones B, C, and D only. Established grass pastures should be overseeded in October-November or February-March. Seedling vigor is better than any other clover and especially well suited for seeding into cool season perennial grass sods.

Fertilizer: If soil pH is below 5.5, lime should be applied. Responsive to phosphorus and potassium.

Seasonal Production: April-October in Zones C and D, March-June in Zone B.

Management: Hay should be cut in early bloom stage. Red clover will provide more grazing than ladino clover during summer. Unlike ladino clover, red clover will not tolerate continuous close grazing over long periods of time.

Pests: Many diseases such as powdery mildew, northern and southern anthracnose, and bean yellow mosaic virus attack red clover. Varieties have been developed with resistance to one or more of these diseases.

Courtesy of Dr. G. W. Evers

Spotted or Southern Burclover

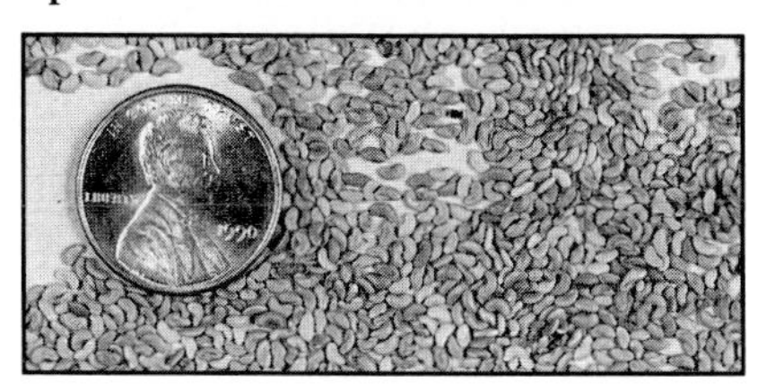

Spotted or Southern Burclover
(Medicago arabica)

Origin: Mediterranean region.

Description: Winter annual with decumbent branches up to 2 feet long. Non-hairy leaves usually with a dark spot in center. Small yellow flowers form small burs.

Primary Adaptation: Zones A and B.

Major Uses: Pasture.

Establishment: Volunteers in many areas. Seed are broadcast at 10 to 15 lb/A in September-October. Seed not readily available.

Fertilization: Grows best in calcareous soils with pH above 6.

Seasonal Production: March and April. Relatively low yield.

Management: Should be quickly utilized due to short season.

Pests: Unknown.

Subterranean Clover

Subterranean Clover
(Trifolium subterraneum)

Origin: Mediterranean region.

Description: Dense low-growing winter annual clover with hairy leaves. Small, mainly white flowers form a small bur which is forced into the surface of the soil. Good reseeding possible. Seed are large and either black or tan.

Primary Adaptation: Zone A and lower part of Zone B. Well-drained soil.

Major Uses: Pasture. Lower-yielding than crimson or arrowleaf.

Establishment: Seed are broadcast at 10 to 20 lb/A. Special inoculant required.

Fertilization: Fairly acid tolerant. Requires adequate phosphorus and potassium for growth.

Seasonal Production: November-December and March-May.

Management: Tolerates close continuous stocking. Tolerates shade better than most clovers.

Pests: Relatively free of pests.

Sweetclover

Hairy Vetch

Sweetclover
Yellow *(Melilotus officinalis)*
White *(Melilotus alba)*

Origin: Europe and central Asia.

Description: Biennial, erect, coarse stems, 4 to 8 feet tall, deep taproot. Yellow or white flowers.

Primary Adaptation: Zones A, B, C, and D. Clay or loam soils. Drought-tolerant and winter-hardy but intolerant of soil acidity.

Major Uses: Pasture. Coumarin in forage reduces palatability to livestock. Low-coumarin varieties are available.

Establishment: Seed can be planted in spring or autumn at 10 to 15 lb/A.

Fertilization: Requires soil pH near neutral, similar to alfalfa.

Seasonal Production: May-August.

Management: New growth comes from stem buds. Overgrazing should be avoided during the first year.

Pests: Sweetclover weevil.

Vetch
Hairy *(Vicia villosa)*
Common *(Vicia sativa)*
Bigflower *(Vicia grandiflora)*

Origin: Mediterranean region.

Description: Viney winter annuals with stems 2 to 4 feet in length. White, purple, or pale yellow flowers are borne in clusters.

Primary Adaptation: Hairy vetch and bigflower vetch in Zones A, B, C, and D, common vetch in Zone A. Vetches need well-drained soil.

Major Uses: Pasture, hay, silage (with small grain companion), green manure.

Establishment: Hairy and bigflower vetch seed are broadcast at 20 to 25 lb/A and common vetch at 30 to 40 lb/A in September-October.

Fertilization: Vetches are tolerant of soil acidity but have a relatively high phosphorus requirement.

Common (Cahaba White) Vetch

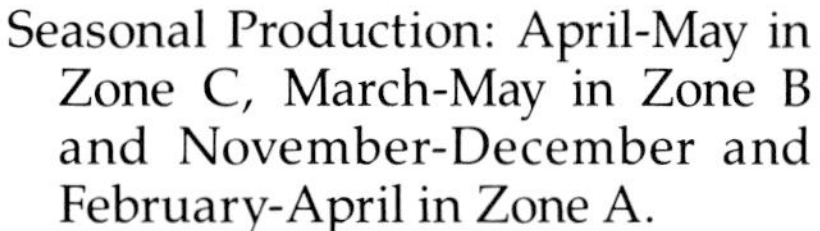

Seasonal Production: April-May in Zone C, March-May in Zone B and November-December and February-April in Zone A.

Management: Grazing should not begin until plants are at least 6 inches tall. Close grazing will destroy buds needed for regrowth. Vetch should be cut for hay in the early bloom stage. A small grain seeded at 60 lb/A makes a good companion crop for vetch grown for hay or silage.

Pests: The vetch bruchid often attacks hairy vetch pods and destroys the seed, reducing reseeding.

Bigflower Vetch

White or Ladino Clover

White or Ladino Clover
(Trifolium repens)

Origin: Mediterranean region.

Description: Fairly long-lived perennial in upper South; short-lived perennial or annual in lower South. Very leafy plants 8 to 12 inches tall that spread by stolons (runners) and form shallow roots at nodes. Leaves are non-hairy and usually marked with a white "V". White flowers are clustered into heads. Seed are extremely small. Intermediate types of white clover can be expected to reseed naturally while giant or ladino types usually do not reseed well in the lower South.

Primary Adaptation: Zones A, B, C, and D. Not productive on droughty soils but will survive considerable dry weather. Tolerant of moderate soil acidity and wetter soils. Grows well in association with cool season perennial grasses, and with dallisgrass, but generally not with bermudagrass and bahiagrass.

Major Use: Pasture. Very high quality grazing plant. Bloat can be a problem.

Establishment: Seed are broadcast at 2 to 3 lb/A with tall fescue, orchardgrass, or Kentucky bluegrass in September-October. Under favorable growing conditions, white clover can shade out seedling grasses so it may be necessary to graze the clover to reduce competition. Established grass pastures can be overseeded in October-November or February-March.

Fertilization: Soil should be limed if pH is below 6. Highly responsive to potassium fertilizer.

Seasonal Production: April-October in Zones C-D, March-June and October-November in Zones A-B if soil moisture is favorable.

Management: Grass competition from undergrazing is one of the major problems in maintaining productive stands of white clover. Grass should be planted in wide rows and clover broadcast to reduce competition. Adequate potassium and phosphorus are important for good production. Grazing should be sufficient to maintain forage height at 1 to 4 inches, preventing shading of clover by the grass.

Pests: A number of leaf and root diseases attack white clover. Close grazing allows light and air penetration to reduce the likelihood of these problems. Virus diseases are the most serious problem and normally plants are infected, weaken, and die after 2 to 3 years. Since no virus-resistant varieties are available, the only solution is to replant the pasture with clover every 2 to 3 years. Various insect pests attack white clover, but no practical control is available.

Winter Pea

Winter Pea or Austrian Winter Pea
(Pisum sativum* subsp. *arvense)

Origin: Mediterranean region

Description: Viney winter annual with stems 2 to 4 feet long.

Primary Adaptation: Zones A, B, and C. Well-drained loam or sandy loam soil.

Major Uses: Silage and green manure. High nutritive quality.

Establishment: Seed are planted in September-October at 30 to 40 lb/A if grown alone or 20 to 30 lb/A if sown with small grain.

Fertilization: Responds well to phosphorus and potassium. Intolerant of highly acid soil.

Seasonal Production: Can be cut for silage in April.

Management: Not well adapted for pasture because plants are easily damaged by trampling. Best used for silage as the crop is difficult to cure for hay.

Pests: Downy mildew can be quite destructive during warm wet winters. Virus diseases can be a problem. Pea weevil, aphids, and nematodes also can cause considerable damage. ■

Chapter 9

Soil Testing and Fertility

PERIODIC SOIL TESTING followed by liming and fertilization according to soil test recommendations is critically important to obtaining good forage production and maintaining forage stands. On most livestock farms in the South, no management practice will have more long-term influence on meat or milk production per acre.

The majority of land producing forage in the South is not regularly soil tested. Furthermore, most of the land which is tested is found to be in need of lime and/or fertilizer, yet much forage land does not receive any annual fertilization (**Table 9.1**). Without question, soil testing and fertilization offer great opportunity for improving forage production within the region.

Soil test recommendations are based on the assumption that the forage produced will be utilized. Thus, it is important for a producer to adjust stocking rates and grazing management as necessary to insure that the forage produced **is** used, either for pasture, haylage or hay. This is essential to obtain the maximum benefit from money spent for fertilizer.

The only way to **know** how much fertilizer or lime is needed to produce high forage yields is to soil test. Perennial pastures should be tested at least once every two to three years. Fields used for hay production, overseeded in winter, or tilled and planted to annual forage crops should be soil tested each year.

The sampling process is the main source of error in soil testing. Therefore, care should be taken to do it correctly. The first time an area is sampled, a single soil sample should represent no more than about 10 acres. On later samplings, one sample may represent larger areas if the entire area is uniform

Table 9.1. Summary of soil test results and forage crop fertilization practices in the South.[1]

Item	Range of responses	Average
Percentage of forage land that is soil tested	5-50%	21%
Percentage of tested land that is below pH 6.0	30-80%	52%
Percentage of tested land that is low in phosphorus	25-70%	46%
Percentage of tested land that is low in potassium	10-60%	33%
Percentage of pasture land receiving fertilizer annually	10-50%	28%
Percentage of hay land receiving fertilizer annually	20-95%	66%

[1]Based on a survey of university agronomists in 10 Southern states conducted in autumn, 1990, by Monroe Rasnake and Garry Lacefield, University of Kentucky.

Courtesy of Dr. Monroe Rasnake

Figure 9.1. The primary source of error in soil testing is the sampling process.

with regard to soil type and has been treated the same.

For each sample, a total of 15 to 20 subsamples should be taken and mixed together, with the sample to be analyzed being taken from the combined subsamples. Subsamples should be taken to plow depth in cultivated fields, but in permanent sods a 2 to 4 inch depth should be adequate.

The Major Nutrients

In forage production, as in the production of most crops, the three major nutrients of concern are nitrogen (N), phosphorus (P), and potassium (K). Each of these must be available in substantial quantities in order to obtain good forage production.

Nitrogen

Nitrogen is the element which normally produces the most dramatic growth response in forage grasses. Nitrogen is a component of chlorophyll, which is critical in the process of photosynthesis. Without photosynthesis, there would be no life.

The Earth's atmosphere is almost 80 percent N, and some N is furnished to the soil as a result of electrical storms and various atmospheric reactions. In addition, N is released by the breakdown of soil organic matter as well as plant and animal residues. However, because N is required by forage grasses in fairly large quantities, and because it leaches rapidly, it should usually be supplied in two or more applications during the growing season of a forage grass. Nitrogen recommendations are based on field research.

Plants have the ability to use N either as ammonium or as nitrate. The ammonium form of N is attracted to, and held by, soil particles, but may be converted to the nitrate form by a microbial process called "nitrification." In this form it can move down through the soil, out of the reach of plant roots.

It is unnecessary and undesirable to apply N to forage legumes. Therefore, it is generally recommended that no N be applied to grass-legume mixtures if the

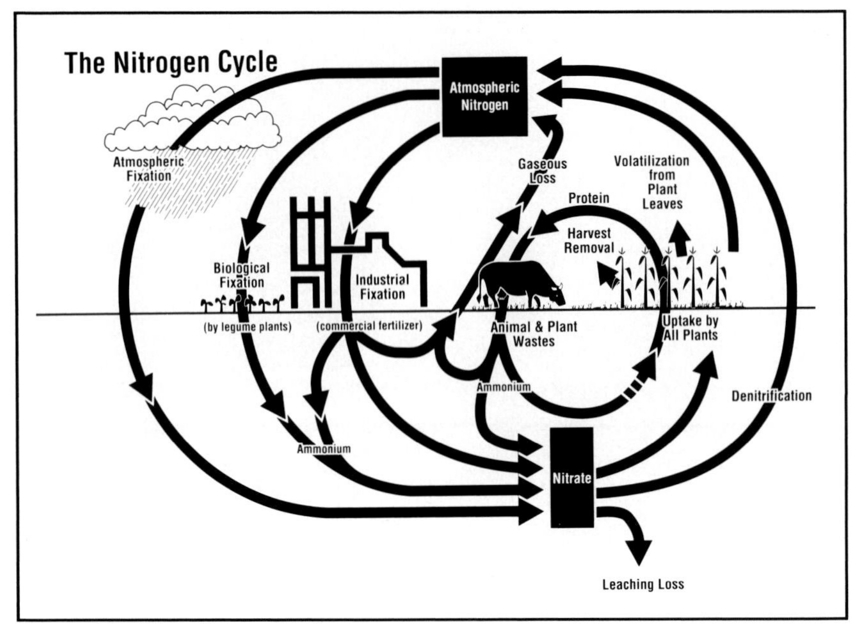

Figure 9.2. The nitrogen cycle.

legume comprises a substantial portion (usually 30 percent or more) of the stand. One exception to this rule is sericea lespedeza/grass mixtures. Although sericea fixes enough N for its own needs, it provides little N for grasses growing in association with it. However, even in this case there is some benefit provided to the grass as a result of N from sericea forage being recycled in dung and urine.

There are several ways by which N applied to grassland can be lost. The most common of these is leaching, which is merely the movement of N by water to depths at which plant roots cannot reach it. Heavy rainfall can cause substantial leaching losses in a short period of time, particularly on sandy soils. Also, N can be removed by water flowing over the soil surface.

Denitrification . . . which results in N being lost into the atmosphere . . . is caused by microorganisms. This sometimes occurs when there are wet soil conditions for an extended period. The potential for denitrification losses is highest on clay soils which have a high water-holding capacity. This type of loss is associated with water-logged soil conditions, which are most likely to occur during the wettest months of the year.

The other type of N loss is volatilization (gaseous losses). This is caused by a chemical reaction which sometimes occurs in surface-applied fertilizers or urine from grazing animals in which N is released into the atmosphere in a gaseous form. The N fertilizer source most likely to volatilize is urea, although in many situations little or no loss will occur even with this product.

Several factors affect the amount of urea volatilization. Temperature has a big influence, making summer applications of this N source risky. Also, because an enzyme in plant tissue promotes the volatilization reaction, the more plant material present, the more likely it is that substantial losses will occur. Volatilization losses are higher and more likely in alkaline soils (pH > 7.0). In general, surface applications of other types of N fertilizer do not usually result in significant volatilization losses.

"Nitrogen cycle" is a term used to encompass all the ways N can be added to, or removed from, the soil. A simplified schematic of this important process is provided in **Figure 9.2.**

Phosphorus

Phosphorus or phosphate (P_2O_5) is important in helping plants manufacture food by using sunlight as a source of energy. It is needed for seed and fruit formation, for proper root growth, and is important to the survival and growth of seedlings. Fortunately, P_2O_5 is not leached from the soil. Phosphate applied but unused one year will generally be available for plant growth in subsequent years, although in some soil types it may be tightly bound and released slowly over a period of years. Phosphate fertilizer can be applied as a single annual application.

Potassium

Potassium or potash (K_2O) is needed by plants in relatively large amounts. This nutrient should be present in adequate amounts at all times, but is particularly important in areas where cold hardiness is a factor in persistence. Even where there is no cold stress, potash is of major importance to stand persistence of legumes.

Potash has also been shown to be important in reducing the susceptibility of bermudagrass to leafspot diseases as well as winterkill, and enhancing root growth and development of many species. With regard to leaching, potash is intermediate between N and P_2O_5. On sandy soils, two applications of potash may be needed annually in order to minimize leaching losses.

Other Nutrients

Several other nutrients are essential for plant growth, but are required in smaller amounts and are less likely to limit crop growth. Calcium (Ca), magnesium (Mg), and sulfur (S) are referred to as secondary nutrients. Soils limed according to soil test should contain adequate Ca. Magnesium is also important in plant nutrition, and can be mixed in other fertilizers or

Figure 9.3. Application of fertilizer according to soil test recommendations is of basic importance in forage production.

be supplied through application of dolomitic lime.

Concern for the environment has resulted in changes in manufacturing processes which have reduced emissions of S into the air and associated contributions of S to the soil in many areas. Sulfur once was an incidental component in many fertilizers, but today most high analysis fertilizers contain no S. Thus, S deficiency is more common than it once was. Recommendations for S vary from state to state, primarily due to differences in soil texture and organic matter content of soils.

The micronutrients include manganese (Mn), iron (Fe), boron (B), copper (Cu), molybdenum (Mo), chloride (Cl), and zinc (Zn). In most areas of the South, and with most Southern forage crops, there is no response to addition of these elements. One exception is B, which is required at a rate of around 1.5 lb/A per year for good clover seed production and at around 3 lb/A annually for alfalfa. Note: A list of the elements known to be essential to plant and animal growth is provided in **Appendix A.27**.

Fertilizers

Soil test results often indicate that the level of one or more nutrients is low. When this is the case, the soil test report normally provides a recommendation to apply enough fertilizer to eliminate the deficiency. The appropriate type and amount of fertilizer to apply will depend on which nutrients are deficient and to what extent.

Fertilizers are sold in many different formulations, usually expressed as ratios. A "ratio" is merely a convenient method of expressing the percentage of N, P_2O_5, and K_2O in a particular type or batch of fertilizer. For example, "triple thirteen" fertilizer is simply a fertilizer having a ratio of N, P_2O_5, and K_2O of 13-13-13. Each 100 lb of the material contains 13 lb of N, 13 lb of P_2O_5, and 13 lb of K_2O.

Table 9.2. Percent N in various fertilizers and amounts of these materials required to supply various levels of N per acre.

Fertilizer material	% N in fertilizer	Pounds of N desired per acre					
		30	60	90	120	150	180
Anhydrous ammonia	82	40	75	110	150	180	220
Urea	46	60	130	200	260	320	390
Ammonium nitrate	33.5	90	180	270	360	450	540
Ammonium sulfate	21	140	280	430	570	710	860
Nitrate of soda[1]	16	190	380	560	750	940	1,120
Calcium nitrate[1]	15	200	400	600	800	1,000	1,200
		Gallons of liquid per acre					
Anhydrous ammonia	82	7	14	21	28	35	42
Solutions	32	9	17	26	34	43	52
	21	13	27	40	53	66	79

[1]May not be readily available commercially.

Table 9.3. Approximate nutrient content of various common phosphate and potash fertilizers.

Material	% Phosphorus (P_2O_5)
Normal superphosphate*	20
Triple superphosphate	45
Monoammonium phosphate	48
Diammonium phosphate	53
Ammoniated superphosphate*	16
Basic slag*	10
Bone meal*	24
Material	**% Potassium (K_2O)**
Muriate of potash	60
Sulfate of potash	50
Sulfate of potash magnesia	22
Potassium nitrate	44
Potassium carbonate*	65
Sodium potassium nitrate*	15
Monopotassium phosphate*	35

*May not be readily available commercially.

Fertilizers such as mentioned in the previous example and which contain N, P_2O_5, and K_2O are referred to as "complete fertilizers." They are used when all three of these nutrients need to be applied. In other cases, soil test reports will indicate that only one or two nutrients are deficient, and therefore a fertilizer which provides only these nutrients is needed. The quantity of nutrients provided by various N fertilizers or by various phosphate and potash fertilizers is provided in **Tables 9.2** and **9.3**, respectively.

Except for nitrate of soda and calcium nitrate, commercially available forms of N tend to increase soil acidity. Most N fertilizers, particularly at high rates, eventually reduce yields by lowering soil pH. Other factors, such as leaching, runoff, and removal of Ca and Mg by crops also tend to increase acidity. The soil pH should be checked by a soil test, and lime applied as needed to neutralize soil acidity.

Figure 9.4. When dung and urine spots become readily apparent, pastures are nutrient deficient.

Table 9.4. Nutrient composition of litter from 147 broiler houses sampled in Alabama, 1977-1987.

	Average analysis, % dry-weight basis	Range, %	Average nutrient content, lb/ton as-is basis
Moisture	19.7	15.0 to 39.0	—
Nitrogen (N)	3.9	2.1 to 6.0	62
Phosphate (P_2O_5)	3.7	1.4 to 8.9	59
Potash (K_2O)	2.5	0.8 to 6.2	40
Calcium (Ca)	2.2	0.8 to 6.1	35
Magnesium (Mg)	0.5	0.2 to 2.1	8
Sulfur (S)	0.4	0.01 to 0.8	6

Average as-is or wet-weight values assume a moisture content of 19.7 percent.
Source: V.W.E. Payne and J.G. Donald, Alabama Agric. Ext. Ser. Cir. ANR-580.

Livestock manures are also a source of nutrients for forage crops. Their use is common in the South and should be encouraged. However, it is important to know . . . as much as is possible . . . nutrient contents of manures being applied, to prevent nutrient imbalances and to minimize the potential for surface loss or leaching of nutrients into ground water. **Table 9.4** shows how variable nutrient content of animal manures can be.

Livestock manure can substantially reduce fertilizer expenses where it is feasible to use it. However, manure is extremely variable, making it difficult to determine whether crop nutrient needs are being met. Prior to using manure as a fertilizer source, it should be analyzed to determine nutrient content. Proper calibration of application equipment is likewise important.

Soil pH

Soil acidity or alkalinity, which is expressed in terms of pH, is also extremely important to crop production. A pH of 7.0 is neutral, while a soil pH below or above this point is acid or alkaline, respectively. Most forage crops commonly grown in the South do best at pH levels between 5.8 and 6.5, although there are exceptions. Generally, nutrients are most available, and many herbicides are most effective, in this pH range. **Figure 9.5** shows effect of pH on availability of some nutrients.

When soil pH becomes too low for good crop growth, as is the tendency for most soils in the South, it becomes necessary to apply lime to the soil to raise soil pH. If lime is needed, the quantity to be applied will be recommended on the soil test report. Dolomitic lime, which contains more Mg than calcitic lime,

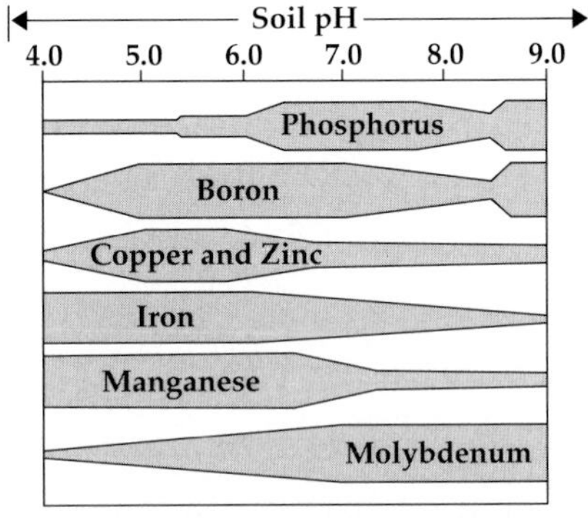

Figure 9.5. The optimum availability of P and the essential micronutrients occurs in the pH range of 6.0 to 7.0.

will be recommended if the Mg level of the soil is low.

Lime moves very slowly through the soil profile. Therefore, it should be tilled into the plow layer in cultivated fields, but in perennial sods, surface application is appropriate. Lime reacts slowly and should normally be applied well in advance of establishing a crop which requires an upward pH adjustment.

In areas where subsoil acidity is a severe problem, 1 to 3 tons of gypsum (calcium sulfate) per acre can be applied. This material **will** move downward through the soil, alleviating aluminum (Al) toxicity, and allowing deeper plant root development.

Using Soil Amendments

Keeping soil pH and fertility at satisfactory levels is important for several reasons. Thousands of agronomic experiments have shown that fertilization increases forage yields. Nitrogen fertilization also affects forage quality, mainly by increasing the protein content of grasses. Finally, proper fertilization promotes the development of healthy, vigorous stands of desirable forage species which offer strong competition to weedy species.

Supplying too much of a nutrient or nutrients can be wasteful or harmful. Plants can "luxury consume" or take up more of an element (particularly K) than needed. In addition, too much of some nutrients will actually **reduce** plant growth. The possibility of over-fertilization emphasizes the need for following soil test recommendations rather than guessing.

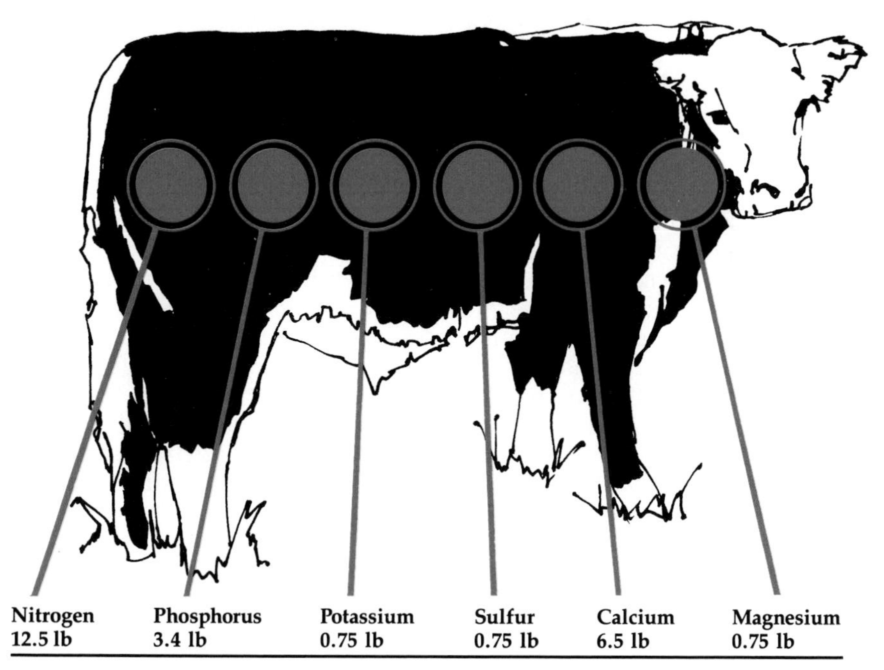

Figure 9.6. Approximate amounts of primary and secondary nutrients present in the body of a 500 lb steer.
Calculations based on percentage estimates from various sources.

Table 9.5. Approximate pounds of nutrients removed by various forage crops at specified yield levels when harvested as hay.[1]

	Species and assumed hay yield, tons/A				
	Alfalfa 5	Tall fescue 3.5	Sorghum-sudan 4	Orchard-grass 3	Bermuda-grass 6
Nitrogen (N)	280	135	160	150	258
Phosphate (P_2O_5)	75	65	61	50	60
Potash (K_2O)	300	185	233	185	288
Magnesium (Mg)	25	13	24	13	18
Sulfur (S)	25	14	–	13	30

[1]Source: Potash & Phosphate Institute

Nutrient Removal and Recycling

A livestock producer should be aware that when forage growth is removed as hay, virtually all the nutrients in the above-ground portions of the plants are being removed. In pasture situations, nutrients in the above-ground portions of plants are consumed by grazing animals, much of which will be recycled through dung and urine. Grazing animals remove relatively small quantities of nutrient elements (**Figure 9.6**).

The effectiveness of the nutrient recycling process in pastures is greatly affected by grazing management. Under non-intensive grazing methods, a high percentage of the nutrients is concentrated in areas where the animals congregate, such as near water, shade, and mineral feeders. Under more intensive grazing methods, such as with high stocking rates and rotational stocking, the distribution of recycled nutrients is much more uniform. Regardless of the grazing method used, losses from volatilization of N from urine range from 30 percent under cool, moist conditions to over 70 percent when hot and dry. Nitrogen is the only element lost in this manner.

Nutrient recycling can have an important influence on the fertility status and fertilizer needs of various fields or even of an entire farm. For example, fields in which livestock are wintered benefit from nutrient recycling much more than fields where livestock are present only for short periods of time. Decisions regarding the buying of hay to be fed to livestock on the farm, or the selling of hay produced on the farm, can greatly affect fertilizer needs. The nutrient flow from, or into, a farm inevitably has an impact on economics.

When the recycling process is poor or forage is removed as hay, certain nutrients may quickly become deficient. **Table 9.5** provides the approximate rate of removal of various nutrients for selected forage crops.

Summary

Providing satisfactory soil fertility and pH for good forage crop growth is one of the keys to successful forage/livestock production. University fertilization recommendations are developed with the idea of producing high forage yield, good forage quality, maintenance of a healthy stand, resistance to encroachment of undesirable species, and economical forage production. Such fertilization recommendations are not necessarily levels which will attain maximum production, but should approach the maximum economical production for proper forage utilization. ■

Chapter 10

Seed

EXCEPT when planting a small area for evaluation purposes, a producer should exclusively plant seed of species and varieties **known** to be adapted and productive in the area(s) in which they will be grown. Agricultural Extension agents can provide variety trial test results. Once species and variety decisions have been made, good quality seed must be located.

Figure 10.1. Use of certified seed helps ensure establishment success.

Seed Quality

Several factors affect the potential of seed to produce a healthy, vigorous stand. Collectively, these factors determine the **quality** of a seed lot. Fortunately, certified seed marketed through normal seed channels are labeled with tags which provide the needed information. The following is normally provided (on a weight basis):

Germination (or Germ)–The percentage of seed which are capable of producing healthy plants when placed in a suitable environment.

Pure Seed–The percentage of seed which are of the variety and species on the label.

Other Crop Seed–The percentage of seed which are of some other crop species.

Weed Seed–The percentage of seed which are of weed species.

Inert Matter–The percentage of sticks, stems, broken seed, sand, or other such material mixed with the seed.

Noxious Weed Seed–Some potentially very harmful weeds are classified as noxious weeds. This refers to the percentage of such weed seed present.

Hard Seed–The percentage of seed which is viable but which will not germinate immediately due to a hard or waxy seed coat. Seed of some forage species, especially some legumes, should be "scarified" before being sold for planting. This refers to a process which seed cleaning facilities use to break down the hard seed coats. Bahiagrass is an example of a forage crop which has a waxy seed coat which requires either time or scarification to allow moisture penetration.

Total Germination and Hard Seed–This is the total percentage of viable seed, obtained by adding together the germination (the percentage of seed which will readily germinate when placed in a suitable environment) and the hard seed (seed which are viable but which will not germinate immediately).

Fungus or Endophyte Level–In some states it is required that the percentage of tall fescue seed infected with the fungal endophyte (*Acremonium coenophialum*) be listed on the seed tag (see Chapter 23).

It is not possible to accurately determine the quality of seed by visual examination. This is why the information on seed tags is of such importance. **All** of this information should be read and understood before seed is purchased.

Classes of Seed Involved in Certification

When a new variety is developed, the plant breeder typically has only a small quantity of seed of the new variety at the date of its "release." The seed must then be increased until there is enough to meet the demand by the public. State Crop Improvement Associations have developed a systematic and efficient means of insuring that the genetic identity of seed is not lost. This is referred to as the Seed Certification Program, and it is used for forage crops as well as other types of crop plants.

Four classes of seed are recognized through state Seed Certification Programs. The first three are: (1) breeder seed; (2) foundation seed; and (3) registered seed. These classifications insure the genetic identity of a variety while seed supplies are being increased.

The fourth class is "certified seed." This term is applied to seed produced in a field which was planted with either foundation or registered seed (registered seed is an optional class). Certified seed is then made available to producers who want to grow the variety. Certified seed is labeled with a blue tag, which serves as a guarantee of genetic purity (**Figure 10.1**).

The seed increase process is closely monitored by Crop Improvement Association personnel and all requirements are strictly enforced. Seed increase fields are inspected to make certain that a designated variety has been planted and to insure that there are no substantial populations of noxious weeds present. These fields are also isolated from other fields of the same crop species to avoid pollination by other varieties or types. The higher the classification (the closer to breeder seed classification), the more rigid the requirements.

Most producers do not need to be concerned with breeder, foundation, or registered seed unless they want to produce, harvest, and sell seed of certified or higher classification. However, forage producers should be familiar with the certified seed classification. In order for seed to be certified, it must meet standards in each of the seed quality areas discussed at the beginning of this chapter. Buying certified seed is the best way to insure that seed purchased is of high quality, genetically pure, and capable of producing thick, productive forage stands.

General Information

Seed is bought and sold on a weight, rather than a volume, basis. Therefore, although a bushel is a commonly used measure for seed, in reality the term "bushel" refers to the number of pounds of a particular crop seed normally required to occupy a bushel volume.

Since the number of pounds per bushel may vary depending on seed production conditions, the term "test weight" is used to refer to

Table 10.1. **Weight per bushel and number of seeds per pound of selected forage crops.**

Forage crop	Approximate lb/bu	Approximate no. of seeds/lb
Alfalfa	60	227,000
Alyce clover	60	301,000
Alsike clover	60	728,000
Annual lespedeza	25	238,000
Annual ryegrass	24	224,000
Arrowleaf clover	60	400,000
Bahiagrass	46	273,000
Ball clover	60	1,000,000
Barley	48	14,000
Bermudagrass, common (hulled)	40	2,071,000
Berseem clover	60	207,000
Bigflower vetch	60	32,000
Birdsfoot trefoil	60	370,000
Black medic	60	266,000
Browntop millet	50	142,000
Button clover	60	153,000
Caleypea (Rough pea)	55	18,000
Carpetgrass	18-36	1,012,000
Corn	56	1,400
Cowpea	60	3,600
Crabgrass	25	825,000
Crimson clover	60	150,000
Dallisgrass	14	281,000
Foxtail millet	50	213,000
Hairy vetch	60	16,000
Indiangrass	—	200,000
Japanese millet	35	143,000
Johnsongrass	28	119,000
Kentucky bluegrass	21	1,440,000
Korean lespedeza (hulled)	59	238,000
Kudzu	54	37,000
Lappa clover	60	680,000
Oats	32	16,000
Orchardgrass	14	416,000
Pearl millet	48	82,000
Persian clover	60	642,000
Prairiegrass (Matua)	10	90,000
Rape	50	156,000
Red clover	60	272,000
Reed canarygrass	47	480,000
Rescuegrass	8-12	52,000
Rose clover	60	164,000
Rye	56	18,000
Sericea lespedeza (hulled)	60	372,000
Small hop clover	60	884,000
Sorghum	50	24,000
Sorghum-sudan	48	35,000
Soybean	60	4,500
Spotted burclover (hulled)	10	250,000
Striate lespedeza (Kobe)	25	200,000
Subterranean clover	60	54,000
Sudangrass	40	43,000
Switchgrass	55	280,000
Tall fescue	25	227,000
Timothy	45	1,152,000
Triticale	48	15,000
Wheat	60	11,000
White clover	60	768,000
White sweetclover	60	259,000

actual weight of a bushel of seed. Test weight is not always specified, but it is commonly used for some types of seed, particularly the small grains. If the test weight is substantially lower than usual, that particular lot of seed is presumed to be of somewhat lower quality than usual.

Seed of various forage crops differ in size, shape, and weight. Thus, the number of seed per pound varies tremendously as does the weight per bushel. Although these numbers are not exact, **Table 10.1** provides approximate weights per bushel and numbers of seed per pound for selected forage crops. Photographs which show the shape and relative size of seed of various forage crops commonly planted or found in the South are provided in Chapters 5 through 8, along with photographs of the forage plants. **Figure 10.2** illustrates parts of legume and grass seed.

Seed should normally not be carried over from year to year. When this is unavoidable, the germination level and vigor of the seed will generally decline. This is particularly true of grasses. If storage is unavoidable, a cool, dry location should be selected. The seed should be retested for germination before use.

Pure live seed–The percentage of pure live seed (PLS) is an indicator of seed quality. It is often used in connection with seeding rate recommendations for species which typically have relatively low germination rates and/or frequently contain a substantial quantity of inert material. As such, PLS is not normally provided on seed tags, but it can be easily calculated if purity and germination are known.

For example, assume that a bag of seed has a purity of 97.5 percent and that the germination is 70 percent. PLS is determined by multiplying the two and dividing by 100. In this case PLS = 97.5 x 70.0 ÷ 100. Therefore, PLS = 68.25 percent.

In order to calculate the amount of seed needed per acre, the PLS seeding rate recommendation should be divided by the PLS percentage and multiplied by 100. Thus, if it is recommended that 10 lb of pure live seed (10 PLS) be planted per acre and the PLS is 68.25 percent as calculated in the previous paragraph, the actual amount of seed needed per acre would be 10 ÷ 68.25 x 100 = 14.65 lb/A. In other words, 14.65 lb of the material taken from the bag or lot need to be planted in order to plant 10 lb of pure live seed. This calculation is only necessary when the seeding rate provided specifies pure live seed or PLS.

Summary

Purchase of certified (rather than non-certified) seed is highly desirable, assuming a suitable variety is available. However, if certified seed is not available, carefully reading the seed tag before purchasing seed is particularly important. Using good quality seed to establish a forage stand does not ensure success, but using poor quality seed practically ensures failure. ■

IDENTIFICATION OF LEGUME AND GRASS SEED PARTS

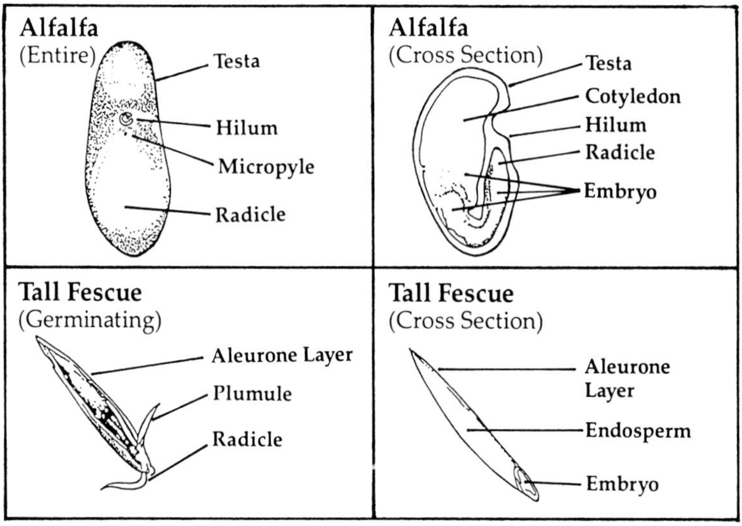

Definitions

Testa–Seed coat.
Aleurone layer–The outer layer of a grass seed.
Hilum–Scar at the point of seed stalk attachment
Micropyle–A pore located near the hilum.
Epicotyl and Hypocotyl–The portions of the embryo above and below the point of cotyledon attachment, respectively.
Plumule–Shoot. End or growing portion of the epicotyl.
Radicle–Root. End or growing portion of the hypocotyl.
Cotyledon–Seed leaf. Grasses have one cotyledon; legumes two.
Endosperm–Food storage area.
Embryo–The tiny dormant plant within a seed.

Figure 10.2. Identification of legume and grass seed parts.

Chapter 11

Forage Crop Establishment

FORAGE CROP ESTABLISHMENT is critical! Errors which will seriously reduce the returns from a forage planting are more likely during establishment than at any other time. The requirements for establishment of the forage crops included in this book vary greatly and space does not permit detailed discussion of each one. However, this chapter emphasizes basic points important in establishment into a prepared (tilled) seedbed, while Chapter 12 discusses sod-seeding. In addition, some points relevant to proper establishment of various species are mentioned in Chapters 5 through 8.

Fertility and Liming

After the decision has been made to establish a forage crop in a particular field, a soil sample should be obtained and submitted for testing to determine nutrient needs. Ideally, this should be done several months in advance of planting. Early testing is particularly important in fields which may require liming to raise the soil pH level.

If a field to be tilled is highly acid, it may be advisable to apply one-half of the recommended lime several months in advance of planting and then to till the field before applying the remainder. This helps mix the lime into the tilled layer so that neutralization of soil acidity is well underway prior to planting.

Tilling the land a few weeks or even months ahead of planting may be helpful even if lime is not incorporated. Subsequent periodic additional tillage prior to planting may stimulate germination of weed seed. Since these weed seedlings will be killed during final seedbed preparation, later weed populations may be reduced.

In addition, rains will settle the soil after tillage and often create a moist, firm seedbed which is more likely to result in a successful forage stand. The potential for erosion should always be considered before making decisions regarding tillage.

The amount of fertilizer recommended by soil test should normally be applied just prior to the final seedbed preparation so that it will be incorporated into the seedbed at a shallow depth.

Seed Considerations

Seed of good quality should always be used (see Chapter 10). Since seed cost is a minor portion of the total establishment expense, use of cheap, low quality seed is not agronomically or economically sound.

Variety tests typically show wide variation in performance among commercially available varieties. Careful review of research information before making variety decisions is time well spent. If the variety desired is not readily available, it may be profitable to go to the effort required to find a source. In some cases, seed of a particular variety can be ordered if the seed dealer has adequate advance notice. If seed is not pre-inoculated, the proper inoculant for legumes should be purchased along with the seed (see Chapter 13).

Seeding Rate. Using a suitable seeding rate is highly important in obtaining a good forage stand.

Many universities and seed companies have developed seeding rate recommendations for various forage crops. **Table 11.1** provides rates often recommended for forages commonly grown in the South. However, these are subject to local variation. Producers should check with their local agricultural Extension agent to obtain recommendations for their area.

Despite the usefulness of general seeding rate recommendations, there can be circumstances which dictate alteration of the seeding rate used. One such factor is seed quality. Most seeding rate recommendations are based on the assumption that germination rates are high. If the seed which is to be used is old or is known to have a low germination level and/or low seed vigor, it is advisable to increase the seeding rate accordingly. However, the best approach is to avoid the use of poor quality seed.

Seedbed preparation has a major influence on the percentage of seed which will ultimately produce healthy plants. A well-pulverized, level seedbed is conducive to planting at the right depth and getting good germination and establishment. The seeding rate recommendations provided in **Table 11.1** are generally appropriate for optimum conditions. It may be desirable to increase seeding rates under less favorable circumstances.

The method of planting is a major factor influencing optimum seeding rate. Generally, a slightly lower seeding rate can be used when drill-planting forage crops than when broadcast-seeding them (this is primarily due to the more precise depth placement which is possible when forage crops are drill-planted). Aerial

Table 11.1. Usual recommended time of planting, seeding rates, and seeding depths for tilled seedbed plantings of forage crops commonly grown in the South.

Forage crop	Time of planting	Seeding rate, lb/A[1]	Seeding depth, inches[1]
Alfalfa	Late summer & autumn, also spring in zones C and D	15-20	1/4-1/2
Alyceclover	Spring, early summer	15-20	1/4-1/2
Bahiagrass	Spring, summer, also autumn in zone A	15-20	1/4-1/2
Barley[2]	Autumn	90-120	1-2
Bermudagrass, common	Spring, summer	5-10 (hulled)	0-1/2
Bermudagrass, hybrid	Early spring	20 or more bu. sprigs	1-3
Big bluestem[3]	Spring	6-10(PLS)[1]	1/4-1/2
Birdsfoot trefoil	Late, summer, autumn	4-6	0-1/4
Black medic	Autumn	10-12	0-1/4
Bluegrass, Kentucky	Autumn	10-15	0-1/4
Caley pea (singletary pea)	Autumn	50-55	1/2-1
Clover, arrowleaf	Autumn	5-10	0-1/2
Clover, ball	Autumn	2-3	0-1/4

(continued)

Table 11.1. (continued)

Forage crop	Time of planting	Seeding rate, lb/A[1]	Seeding depth, inches[1]
Clover, berseem	Autumn	20-25	¼-¼
Clover, crimson	Autumn	20-30	¼-½
Clover, red	Autumn, & early spring zones C & D	12-15	¼-½
Clover, subterranean	Autumn	10-20	¼-½
Clover, white & ladino	Autumn, & early spring zones C & D	2-3	0-¼
Cowpea	Late spring, early summer	100-120	1-3
Dallisgrass[3]	Spring	10-15(PLS)[1]	¼-½
Fescue, tall	Fall, early spring in zones C & D	20-25	¼-½
Indiangrass[3]	Spring	6-10(PLS)[1]	¼-½
Johnsongrass	Late spring, early summer	20-30	½-1
Lespedeza, annual	Early spring	25-35	¼-½
Lespedeza, sericea	Spring	20-30	¼-½
Millet, browntop	Mid spring, early summer	25-30	½-1
Millet, pearl	Mid spring, early summer	25-30	½-1½
Oats[2]	Autumn	90-120	1-2
Orchardgrass	Autumn	15-20	¼-½
Reed canarygrass	Autumn	5-8	¼-½
Rye[2]	Autumn	90-120	1-2
Ryegrass	Autumn	20-30	0-½
Sorghum, forage	Late spring, early summer	15-20	1-2
Sorghum sudan hybrids	Late spring, early summer	30-35	1-2
Sudangrass	Late spring, early summer	30-40	½-1
Switchgrass[3]	Spring	5-6(PLS)[1]	¼-½
Timothy	Late summer, early autumn	6-8	¼-½
Triticale[2]	Autumn	90-120	1-2
Vetch, common	Autumn	30-40	1-2
Vetch, hairy	Autumn	20-25	1-2
Wheat[2]	Autumn	90-120	1-2

[1]These are general guidelines for broadcast plantings. Drill plantings may require less seed. Specific recommendations for a given area may vary and some crops are adapted only in certain regions (see Chapters 5-8). Seeding rates may vary considerably from these when these crops are planted for seed. PLS = Pure live seed (see Chapter 10).

[2]The small grains usually have approximately the following lb/bushel: barley-48; oats-32; rye-56; wheat-60; and triticale-48. See Table 10.1, page 85.

[3]Germination rate is often low.

seeding from a plane or helicopter may justify using up to twice the usual amount of seed. A higher-than-normal seeding rate may also be justified when a soil-conserving cover is needed quickly to protect highly erodible areas.

Since seed cost is generally one of the lowest inputs in the establishment of a forage stand, producers should use **at least as much** seed as is recommended. No one wants to waste money on seed, but it is better to exceed the recommended seeding rate when establishing forage crops than to use too little seed. This is particularly true if planting conditions are less than optimum. When planting mixtures of forage species, it is advisable to use one-half to two-thirds as much seed of each species as would be used when establishing only one species.

Time of Planting

Planting dates vary considerably depending on location, but general recommendations are also presented in **Table 11.1.** Weather plays a major role in determining the proper time to plant, and there is usually a fairly short period of time during which planting conditions are optimum or near optimum. To miss this period can mean the difference between success and failure.

Adverse weather conditions often delay plantings. Therefore, it is advisable to plant a crop as early as possible within the recommended period. Seeding an autumn-planted crop too late may result in winter-kill. Planting a spring-planted crop too late may result in excessive weed competition and possibly death of forage seedlings due to drought. Conversely, seeding an autumn-planted crop too early may cause plant death due to heat stress, while planting a spring-planted crop too early may result in cold damage.

Final Seedbed Preparation

In addition to applying and incorporating needed lime and fertilizer, the seedbed should be relatively firm. This is especially important for small-seeded legumes such as alfalfa and clovers. Weeds and plant residue should be eliminated from the soil surface, and the seedbed should be reasonably smooth. Ridges and depressions need to be minimized as they may hinder the operation of equipment for years to come. When establishing hay meadows, care should be taken to insure obtaining as smooth a field as possible to reduce costly equipment repairs and to facilitate efficient harvesting.

Weed Control

Weed control needs vary tremendously among forage crops. It may be desirable to apply herbicides before planting, at planting time, or after emergence of a forage stand, depending on the crop being established and the weeds present or anticipated. See Chapter 15 for additional discussion of weeds in forage crops and consult local recommendations for weed control in specific forage crops.

Insect Control

Problems with insect pests vary greatly with different crops and different situations. However, it is desirable to closely observe a new forage stand for insect damage. If such damage exists, the insect should be identified and a treatment threshold level should be determined. Agricultural Extension agents can assist in this regard.There are usually several insecticides which can be used for

Figure 11.1. Forage crops can be seeded effectively by several different methods.

control of a pest. Insect control is especially important during establishment because mature, well-established forage plants with a well-developed root system can withstand considerable damage, while seedlings cannot.

Seeding Technique

Forage crops can be seeded by many different techniques, **Figure 11.1**. Spin-seeders which operate from the power takeoff of tractors or from batteries, band seeders, cultipacker seeders, and drills can all be successfully used. In general, large-seeded species such as the small grains or vetch are best seeded with a drill.

Broadcasting followed by shallow disking may cover seed too deeply. Small-seeded species may be drilled if care is taken to avoid planting too deeply. A cultipacker-seeder is the ideal tool to use for planting clovers, birdsfoot trefoil, and alfalfa.

Regardless of the equipment used, a record should be kept of the drill setting used to plant a certain rate of seed. Such information can be of great value the next time the drill is used to plant the same forage species.

Broadcast seeding is also acceptable for planting small-seeded forage crops, if followed by adequate firming with a cultipacker to obtain good seed-soil contact. This is an important point. Many stand failures of clovers and alfalfa have resulted from not having a properly firmed seedbed. If the seedbed is not firm, the seed may end up too deep due to settling of the soil. A "rule of thumb" for evaluating a seedbed for these crops is that when one walks over the field, the shoe prints should not be much deeper than the sole of the shoe.

Both the seed and the inoculant (in the case of legume plantings) need to survive the planting process. Legume seedings should ideally be in moist soil to insure survival of the inoculant and to enhance germination. Legume seed should generally not be mixed with fertilizer as salts can kill legume inoculant. It is desirable to have a shallow soil cover over legume seed to protect the inoculum from the elements and to reduce risk for the germinating seed. However, great care must be taken to prevent seed from being covered too deeply, or they will not emerge (Figure 4.3). Another "rule of thumb" is that seed should be planted no deeper than 8 times the thickness of the seed. General guidelines regarding seeding depth are provided in **Table 11.1.**

Most farms contain a diversity of soil types, so it is important to match the crop to the soil on which it will be planted. When a producer does not know what forage species are adapted to a particular area, a soils map, or a knowledgeable and experienced individual such as an Extension agent, should be consulted.

Despite all efforts, failures sometimes occur in forage stand establishment. When a failure does occur, it is important to diagnose the cause, so the problem can be avoided in the future. A checklist of potential problems which can be responsible for stand failures is provided in **Table 11.2.**

Table 11.2. A checklist of potential problems in forage establishment.

Failure to germinate
Dry seedbed
Non-viable seed
Hard or dormant seed
Unfavorable temperature
Herbicide residue
Water logged soil
Seed germinated but did not emerge from soil
Planted too deep
Soil crusted at surface
Poor seedling vigor
Insects or diseases
Extreme temperature–too cold or too hot
Seedling emerged but did not survive
Soil too acid or low fertility
Insects or diseases
Drought
Weed competition
No legume nodulation
Winterkill
Grazing too early–pulled up plant

Vegetative Establishment

Some forages, such as hybrid bermudagrasses and perennial peanuts, are routinely established from vegetative plant parts rather than from seed. These crops produce little or no viable seed and **must** be vegetatively propagated.

One of the first and most important considerations in establishing hybrid bermudagrass fields is obtaining high quality certified planting material. Sprigs are usually sold by the bushel, but could also be offered by the cubic foot or by the thousand. A bushel normally contains about 1,000 sprigs, while a cubic foot will contain around 800.

Some sprig producers will furnish sprigs and plant them in a prepared seedbed. If a livestock producer wants to obtain and personally plant sprigs, great care needs to be taken to handle them correctly before planting. If possible, sprigs should be planted the same day they are dug. They should be kept moist and, if necessary, turned often to prevent heating. When transported, they should be kept tightly covered to prevent drying by sun and wind.

Nursery Establishment. Many livestock producers who plan to establish sizable acreages of bermudagrass establish a nursery in which they grow their own planting material. This greatly reduces the cost of establishment and also provides a convenient source of material.

A bermudagrass nursery should be established on well-drained, preferably sandy, soil that is free of perennial grasses, especially common bermudagrass. A good seedbed should be prepared and fertilizer should be applied according to soil test recommendations. To avoid excessive stimulation of annual grasses, only around 30 lb/A of nitrogen (N) should be applied at planting.

To reduce weed competition in row plantings, the fertilizer can also be applied at appropriately reduced rates in rows 3 to 4 feet apart. Then, **when the soil is moist,** bermudagrass sprigs can be planted in the rows where the fertilizer was applied. Individual plants should usually be spaced at 18 to 24 inches within rows.

Whether bermudagrass is planted in rows or broadcast, application of a preemergence herbicide to inhibit crabgrass and other annual weeds is essential. The bermudagrass sprigs should have roots covered with soil, but ideally the tips of the sprigs should be at or near the soil surface. In moist soil, a 2 to 3 inch planting depth is about right. Slightly deeper planting may be appropriate if moisture is marginal. If necessary, tillage or direct applications of herbicides can be used to control weeds between the rows.

Additional N can be applied to the rows within 3 to 4 weeks after the grass has begun to root and grow. Once the rows begin to lap, or if bermudagrass is planted broadcast rather than in rows, 40 to 60 lb/A of N should be broadcast every 4 to 6 weeks until September.

Field Planting Techniques. Weed control should be started well ahead of planting. Any existing perennial plants, especially common bermudagrass, should be killed. If practical, the land should be tilled in the autumn prior to planting.

Bermudagrass sprigs can be planted in several ways. It is possible to plant by hand, but the labor involved makes this approach prohibitive for most producers. However, if planted in rows by hand, as little as 10 bu/A can be used.

When bermudagrass is planted with a commercial sprigging machine it is customary to use at least 20 bu/A. Up to 40 bu/A might be appropriate for a slow-spreading variety. Most states in the lower South have custom operators who will sell sprigs and/or plant fields. Since these people are experienced in planting hybrid bermudas and they generally have the proper equipment to do a good job, hiring them is usually money well spent.

A producer who has established a nursery must devise a way to dig and plant sprigs. If there is no access to a sprig digger, a spring toothed harrow can be used to loosen sprigs, followed by raking them together with a hay rake. Then they can simply be forked or scooped onto a truck or trailer and broadcast over a fresh, moist seedbed. Next they should be lightly disked into the soil followed by firming the field with a cultipacker. At least 20 to 30 bu/A should be

Figure 11.2. Sprigging of bermudagrass (right) should be followed by rolling or cultipacking (left).

used when this technique is employed.

Bermudagrass should be planted into a moist, freshly prepared seedbed and a preemergence herbicide should be applied just after planting. In addition, a postemergence herbicide application may be needed to control broadleaf weeds. If annual grasses are a problem with a new planting, frequent clipping will help reduce competition. Keeping fertility high is essential in getting good spread of the bermudagrass.

The timing of planting is also of critical importance in getting a stand of bermudagrass. The best time to plant is just before the bermudagrass rhizomes break dormancy. In most of the South this is in late February or March. The longer planting is delayed after this time, the lower the chances of success.

Failures of bermudagrass establishment with sprigs are usually due to:

(1) Planting when there is inadequate soil moisture;
(2) Using poor quality sprigs (usually dried out);
(3) Planting too few sprigs;
(4) Covering sprigs too deeply;
(5) Failing to firm soil around sprigs;
(6) Failing to control weeds.

Establishment from Clippings

Clippings (the above-ground growth) of hybrid bermudagrasses can also be used to establish a new stand, but the feasibility of this varies greatly with variety as some root much more easily than others. With appropriate varieties, clippings to be used for establishment can be cut when forage growth reaches about 18 inches. The clippings should be generously spread over freshly-tilled **moist** soil and lightly disked in the same day they are cut. The objective is to cover one or more nodes of the stems, on average. This technique is more labor intensive and risky than machine sprigging, but is inexpensive. ■

Chapter 12

Sodseeding/No-Tillage Planting

SODSEEDING is a general term used to describe the practice of establishing forage crops into perennial, usually grass-dominant, hay and/or pasture fields without destroying the existing sod. Many other terms are also used to describe this practice including: renovation, overseeding, interseeding, frost seeding, honeycomb freeze seeding, no-till seeding, and minimum-till seeding, as well as tread-in, trampling and walk-on seeding.

Applications of this practice have been used with success throughout the South for many years. However, two situations are by far the most common. In the upper South, establishment of legumes (alfalfa, red clover, and/or white clover) into tall fescue, Kentucky bluegrass or orchardgrass sod is an important forage improvement practice (**Figure 12.1**).

In the lower South, cool season annual grasses and/or legumes are often planted into dormant perennial warm season grass sods, especially bahiagrass or bermudagrass. These and other sodseeding techniques can be used to great advantage by many Southern livestock producers.

Benefits of Multiple-Species Pastures and Hayfields

There are several important potential benefits associated with establishing other forage crops into existing grass sods. Research and farmer experience have shown that

Figure 12.1. Red clover established into tall fescue pasture. Red clover and/or white clover (inset photo) is commonly sodseeded into tall fescue, orchardgrass, or dallisgrass.

the addition of legumes to grasses can result in improvements such as:

(1) Higher yields. The total yield of forage per acre is likely to be increased unless high levels of nitrogen (N) fertilizer are used on grasses. A study conducted in Kentucky compared renovation of a tall fescue pasture using red clover to fertilizing the grass with N (**Table 12.1**). In this test, red clover growing with tall fescue produced higher yields than tall fescue fertilized with up to 180 lb N/A.

(2) Improved quality. Adding legumes to grass fields improves forage quality over grass alone. This results in increased palatability, intake, digestibility, nutrient content, and animal performance.

Research has shown that legumes improve animal growth rates, reproductive efficiency and milk production. For example, growth rates of beef steers in Alabama were increased when ladino clover was planted in endophyte-infected tall fescue pastures (**Table 12.2**). Higher gains per acre were also recorded, mainly as a result of improved forage quality. In another Alabama study, overseeding bermudagrass

Table 12.1. Dry matter yields of tall fescue-red clover vs tall fescue with N fertilizer. (Lexington, Ky, 2-year average).

Treatments	Yield, lb/A
Tall fescue-red clover 6 lb seed/A	11,100
Tall fescue + nitrogen	
0 lb N/A	3,900
90 lb N/A	6,700
180 lb N/A	9,900

Source: T.H. Taylor. 1988. Kentucky Agric. Ext. Ser. Pub. AGR-26.

Table 12.2. Performance of steers grazing endophyte-infected tall fescue with and without ladino clover in North Alabama, 2-year average.

Pastures	Average daily gains, lb	Gain/ steer, lb	Gain, lb/A
Tall fescue + ladino clover	1.53	307	582
Tall fescue + 150 lb N/A	1.06	203	374

Source: C.S. Hoveland, R.R. Harris, E.E. Thomas, E.M. Clark, J.A. McGuire, J.T. Eason and M.R. Ruf. 1981. Alabama Agric. Exp. Stn. Bull. 530.

Table 12.3. Performance of beef cows and calves on Coastal bermudagrass pasture overseeded with winter annuals in southeastern Alabama, 3-year average.

			Cows	Calves	
Species overseeded on bermuda sod	N rate, lb/A/yr	Dates on pasture	Average daily gain, lb	Average daily gain, lb	Gain, lb/A
Rye & arrowleaf & crimson clovers	100	Jan. 8-Oct. 5	0.90	1.91	511
Arrowleaf & crimson clovers	0	March 11-Oct. 5	1.37	1.94	410
Ryegrass	150	Feb. 14-Oct. 5	0.81	1.76	422
None	100	April 6-Oct. 5	0.49	1.57	293

Source: C.S. Hoveland, W.B. Anthony, J.A. McGuire and J.C. Starling. 1977. Auburn Univ. Agric. Exp. Stn. Bull. 496.

Table 12.4. Value and amount of N fixed by various legumes.

Crop	N fixed, lb/A/year	N value, $, @ 25¢/lb	N value, $, @ 35¢/lb
Alfalfa	150-250	38-63	53-88
Red clover	75-200	19-50	26-70
Ladino clover	75-150	19-38	26-53
Vetch, lespedeza and other annual forage legumes	50-150	13-38	18-53

with arrowleaf and crimson clovers increased both cow and calf average daily gains with no N fertilizer (**Table 12.3**).

(3) Nitrogen fixation. Legumes fix N which can be used for growth by the legumes and also by associated grasses (see Chapter 13). The value of the N fixed by legumes depends on the cost of N fertilizer. The values in the left column of **Table 12.4** are based on N priced at 25 cents/lb or ammonium nitrate fertilizer at $170/ton.

(4) Longer growing season. Most cool-season grass growth occurs during spring and autumn. In the upper South, legumes such as red clover, alfalfa or lespedeza make more summer growth than cool-season grasses. Thus, growing grasses and legumes together can improve the seasonal distribution of a pasture and provide more summer growth.

Addition of cool season annuals to bermudagrass, bahiagrass or dallisgrass permits production of quality feed during winter and early spring when pastures would otherwise be unproductive. Many cool season annual forages can be sodseeded, either alone or in mixtures, but annual ryegrass is the single most widely-used species. In addition, annual clovers such as arrowleaf or crimson clovers are often used, as are rye and wheat. Seeding cool season species into warm season perennial grass sods can substantially lengthen the grazing season.

Other systems used in the South involve the seeding of various forage species into crop stubble with no-till techniques. Examples include seeding alfalfa or other species into warm season annual grass stubble (sorghum-sudan hybrids and millets), corn silage stubble, and small grain. Large-seeded species such as small grains must be drilled into warm-season grass sods. Small seeded species such as ryegrass and clover may be drilled or broadcast.

Sodseeding Principles

In many situations, sodseeding offers advantages over tillage for forage establishment. Since much forage land throughout the South is rolling and prone to erosion when tilled, sodseeding can conserve soil (see Chapter 32). Likewise, sodseeding can conserve time, labor, fuel and soil moisture.

Soil Conservation. Following heavy rain on a well-tilled seedbed, severe soil erosion may be readily apparent. Only after the plants are established is the amount of run-off water and soil loss greatly reduced under tillage establishment. Sodseeding can minimize soil loss during establishment.

Time. Conventional seedings normally require several trips over the field to get the seedbed properly prepared and seed precisely planted.

Sodseeding techniques may require only one pass over the field. Even if additional trips are required for fertilizing and spraying, total trips are still fewer than required by conventional seeding techniques.

Scientists in Kentucky studied the time required to establish clover into tall fescue pastures using a sodseeder as compared to disking and seeding conventionally. Their results showed the land area which could be covered per unit of time was 3.90 acres per hour using a sodseeder and only 0.62 acres per hour using conventional techniques.

Fuel. If fewer trips are required over a field to accomplish the same results, fuel will also be saved. In the studies mentioned above, conventional seedings required 4.95 gallons of fuel per acre while the sodseeding method required only 1.04 gallons.

Moisture. Moisture conditions in either spring or late summer can potentially show an advantage for sodseeding. In late winter or spring, moisture is often surplus, delaying seedbed preparation and seeding if conventional (plow-disk) techniques are to be used. With sodseeding techniques, seedings can be made earlier since a no-till drill can be used in a field several days or even weeks earlier than it can be tilled.

During late summer and early autumn, moisture shortage is common in the South. The limited soil moisture is often lost during seedbed preparation with conventional seeding. With sodseeding, very little moisture is lost since only a small amount of soil is disturbed.

Specific recommendations will not be given to cover all possible sodseeding situations existing throughout the South. However, some general principles include:

(1) **Fertility.** A soil test should be taken and used as the guide for amounts and kinds of soil amendments to be applied. Application of any needed lime, phosphorus (P) and potassium (K) is especially important for successful

Figure 12.2. Cattle grazing tall fescue following clover seeding, near Owensboro, Kentucky.

Figure 12.3. Experimental studies showing competition from existing grass controlled by tillage, herbicides and mowing.

establishment of legumes into grass sods. Nitrogen should be applied whenever pure stands of winter annual grasses are planted. Rates can be reduced when legumes are seeded with annual grasses, depending on seeding date and species or species combination used.

(2) **Reduce existing vegetation.** Failure to control existing grass vegetation is the most common reason for stand failures from sodseeding. Reducing the vegetative cover on the field is **necessary** for successful sodseeding. Reducing vegetative cover will reduce competition for the newly seeded forage species as well as make it easier to get good seed-soil contact during seeding. Existing grass vegetation can be reduced by heavy grazing, herbicides, tillage, or combinations of these approaches. Methods used for reducing vegetation will depend on several factors, including amount and kind of grass, legume(s) to be seeded, livestock numbers, and experience.

Many producers use their livestock as a tool for improvement of tall fescue pastures. Heavy stocking will reduce the existing vegetation, and hoof action will provide some "seedbed" preparation. In late winter, red, ladino or white clover seed can be broadcast and "trampled in" with excellent results.

Clover seed placed in contact with bare ground usually gets covered at or near ideal depth after a few freeze-thaws. Late freezes which give a honeycomb appearance provide an excellent surface for legume seed. As the "honeycomb" thaws, the small seed usually settle just below the soil surface in an excellent environment for later germination.

Herbicides can be used to kill the entire stand of existing

Figure 12.4. Establishing alfalfa via no-tillage techniques can result in excellent stands when proven principles are followed and weather conditions are favorable.

grass (i.e., no-till alfalfa seeding) or to suppress a sufficient amount of grass to get the newly seeded legumes established. The "zebra technique", sometimes used when establishing clover in cool season grass sods, requires adjusting the nozzles on the spray boom such that only 50 percent of the grass is sprayed. After the herbicide has suppressed the grass, the field gives a striped or "zebra" appearance with half of the grass brown and the remainder green. Clover seed can be planted either simultaneous with or following spraying.

(3) **Use high quality seed.** Regardless of the species or combination of species to be seeded into existing grass sod, the use of high quality seed is essential. **Poor quality seed is never a bargain, regardless of price.** High quality certified seed of an adapted variety should be used when available.

(4) **Inoculate legumes.** If not pre-inoculated, all legume seed should be inoculated properly with the correct strain of bacteria before seeding (**Appendix A.31.**).

(5) **Plant on time with the correct amount of seed.** Seeding dates and rates vary across the South. Local Extension offices can provide recommendations for specific areas.

(6) **Seed-soil contact.** Many methods exist for sodseeding (**Figures 12.5 and 12.6**), ranging from cyclone seeders to no-till drills. Regardless of the method used, the objective is to plant high quality seed uniformly throughout the field.

(7) **Control competition.** Competition from existing vegetation must be suppressed

Figure 12.5. Proper techniques of overseeding a warm season grass sod resulted in this ryegrass/arrowleaf clover pasture near Montgomery, Alabama.

Figure 12.6. Ladino clover being overseeded on closely grazed tall fescue pasture, near Princeton, Kentucky.

prior to and at seeding. Grazing or mowing can be used to keep the existing grass short during the first few weeks after planting while the newly-seeded plants are becoming established.

(8) **Control pests.** Insect populations should be monitored and controlled as necessary during the establishment phase. Safe, effective insecticides are available and can be used if sufficient insect damage is apparent, **Figure 12.7.**

Table 12.5 summarizes no-tillage forage systems. Sodseeding offers much potential for producers who are willing to exercise a higher level of management.■

Figure 12.7. Arrowleaf clover established in sod plots in Georgia. At left, insecticide was used to control striped field crickets. At right, no insecticide was used.

Table 12.5. Summary of no-tillage forage systems.

Sod or crop stubble to be planted into	Species planted	Zone	Usual planting time
Cool season perennial grasses			
Tall fescue,	Alfalfa	A,B	Autumn
orchardgrass,		C,D	Late win.-early spr.,
Kentucky bluegrass,			Late sum.-early aut.
timothy	Red clover	A,B	Late win.-early spr.,
			Late sum.-early aut.
		C,D	Late win.-early spr.,
	Birdsfoot trefoil	B	Autumn
		C,D	Late win.-early spr.,
			Late sum.-early aut.
	Ladino/white clover	ABCD	Late win.-early spr.,
			Late sum.-early aut.
Warm season perennial grasses			
Bermudagrass and bahiagrass	Small grains; ryegrass; arrowleaf, berseem, crimson, and subterranean clovers; vetch	A,B	After killing frost
Dallisgrass	White and red clovers, caleypea	A,B	After killing frost
Small grain stubble	Pearl millet, sorghum-sudangrass	C,D	Late spring
Rye, wheat, oats			
	Alfalfa	C,D	Late sum.-early aut.
	Red clover	C,D	Late win.-early spr.

Chapter 13

Legume Inoculation

NITROGEN (N) IS REQUIRED in large quantities for high levels of forage production, and N deficiency is a major limitation to forage/livestock production. The atmosphere consists of about 79 percent N and each acre of land has an estimated 35,000 tons of N above it. However, the N in the atmosphere is in the form of an inert gas which is unavailable to plants.

There are several sources of N for plant growth. Minute quantities are supplied by free-living soil bacteria and blue-green algae. Rainfall typically adds a few pounds per acre, mainly converted from atmospheric N by lightning. Most commercial N fertilizer is produced by the Haber-Bosch process which, under conditions of high temperature and pressure, utilizes natural gas for the production of ammonia. A final source is legume plants, which produce substantial quantities of low-cost N.

The unique ability to obtain N from the air makes legumes especially valuable in forage programs. The N they produce can be used for their own growth, by associated grasses, or by crops in rotation. In addition, pasture legumes can ultimately provide N for plants a considerable distance from where they are growing because much of the N taken into the bodies of livestock is recycled in dung and urine (See Chapter 9).

Most legumes have a symbiotic (mutually beneficial) relationship with bacteria in the genus *Rhizobium*. The bacteria infect the roots of legume plants from which they obtain food,

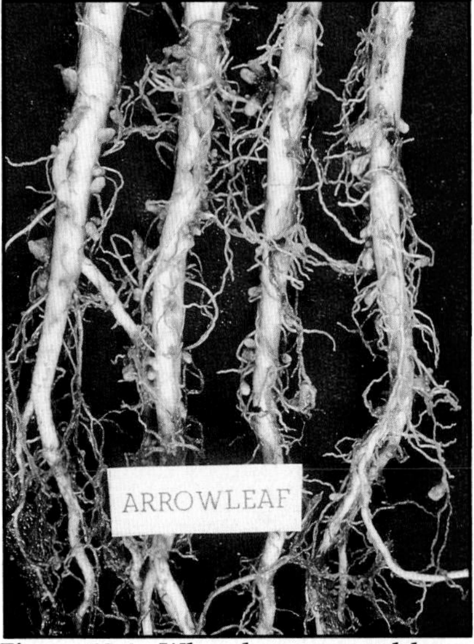

Figure 13.1. When legume seed have been properly inoculated, N is fixed in nodules on the plant roots.

and the bacteria obtain N from soil air and "fix" it in a form usable by the plants. The N is accumulated in small appendages called "nodules" which form on legume roots.

The enzyme nitrogenase is produced by the bacteria and it is this compound which, in the presence of molybdenum, is responsible for reducing gaseous N to ammonia. The nitrogenase enzyme is the only natural substance known to react with inert atmospheric N at normal temperature and pressure.

Nitrogen fixation is influenced by many factors including *Rhizobium* population in the soil, temperature (approximately 77° to 84° F is optimum), soil moisture (which can be either inadequate or excessive), soil pH, soil fertility, amount of shading, and legume species. Nitrogen fixation is adversely affected by factors which reduce photosynthesis. Thus, N fixation is higher during the day than at night, during sunny rather than cloudy weather, and when leaf area is high as compared to when plants have been heavily defoliated.

Several nutrient elements deserve special mention in connection with N fixation. Calcium and/or potassium (K) deficiency decreases N fixation, while application of more than around 25 lb N/A will generally reduce legume N fixation and stimulate the growth of competing plants. Competition from weeds or grasses reduces N fixation by competing with legumes for K and by shading. *Rhizobium* bacteria also require minute quantities of molybdenum and N fixation will be hindered where supplies are inadequate.

Good N fixation requires that large numbers of live *Rhizobium* bacteria are present on legume roots. Assurance that this will occur can be obtained by "inoculating" the seed (putting bacteria on the seed), a standard practice since the early 1900s.

It is important to distinguish legume seed inoculants from other products marketed as soil inoculants or "activators" of microbial processes. Inoculation of legume seed with *Rhizobium* bacteria is of proven practical value and all references made here to inoculation refer *only* to this process.

Specific species of *Rhizobium* bacteria are required for specific species of legumes (see **Appendix A.31**), and since some *Rhizobium* strains are more effective than others, inoculant companies select within species. Thus, use of high quality inoculum of the proper type is critically important. Peat-based legume inoculants are commercially available where legume seed is sold; usually one package can be used to inoculate 50 lb of seed.

When inoculant is purchased, a producer should check the label to be certain that: (1) the legume to be inoculated is specifically listed on the package; and (2)that the expiration date has not been exceeded. Prior to use, the inoculant should be stored in a reasonably cool place (40° to 70° F or refrigerated) and out of direct sunlight. Countless inoculation failures have resulted from inoculant being carelessly left in the heat of a pickup truck cab.

The native population of *Rhizobium* may vary from none to several million per gram of soil. It is of critical importance to inoculate legume seed when planting in areas where a legume species has not been grown before. Some legume bacteria will remain in the soil for several years, but since inoculant is inexpensive, inoculation is cheap insurance even where legumes have previously been grown. Inoculation should always be done with fresh inoculant, not with inoculant saved from a previous year. It is preferable to avoid mixing inoculated seed with fertilizer because fertilizer

Figure 13.2. The inoculant package should indicate what species of legumes can be inoculated, and should provide an expiration date for use.

salts can kill *Rhizobium* bacteria, especially uncoated seed, under conditions of high humidity or if the seed is not planted immediately.

Legume seed should be inoculated just prior to planting using the following procedure.

Inoculation Process

Step 1. Place seed in a tub or similar open-topped container.

Step 2. Slightly dampen seed with a syrup-water mixture, or a commercial sticker solution. (Water alone is not satisfactory because the inoculum does not stick to the seed once it has dried.) Stir seed thoroughly to insure all seed are damp. Commercial stickers can improve adhesion of the inoculum to the seed and prolong survival of the bacteria. (Caution–It is important to use only the recommended amount of commercial sticker or a small quantity of syrup-water mixture; otherwise the seed may stick together.)

Step 3. Add inoculum and stir thoroughly to evenly distribute. If the seed are too sticky to easily flow through planting equipment, finely ground agricultural lime (**not** pickling lime) can be mixed with it to solve the problem. Inoculated seed should be protected from the sun and heat as much as possible and should be planted within a few hours after inoculation. If planting is not possible on the day the seed are first inoculated, it is advisable to inoculate again.

Low soil pH (less than about 5.8), or extreme heat or drought can quickly reduce *Rhizobium* bacteria numbers. Under extremely adverse conditions less than 0.01 percent of the original *Rhizobium* population present on the seed will survive for two days. Bacteria survival will be much better if the seed are planted in a slightly moist soil or if planted just prior to a rain. Although great care should be taken to avoid covering small-seeded legumes with more than 1/4-inch to 1/2-inch of soil, a light or partial soil covering, such as can be accomplished with use of a cultipacker-seeder, will help protect the bacteria.

Seed of alfalfa, red clover and some other major annual and perennial legume species are often pre-inoculated and have a protective coat requiring no further inoculation. Sometimes fungicides, insecticides, lime, or minor elements are also incorporated into the seed coats of pre-inoculated seed. Pre-inoculated seed normally maintain very high levels of bacteria which are protected by the seed coating better than farm-inoculated seed. Seedling survival for such seed is often higher, thus allowing slightly lower seeding rates. Pre-inoculated seed should be kept in a cool, dry place until planting time in order to protect the bacteria on the seed surface and should be planted within a few months.

Rhizobium bacteria infect root hairs of legume plants near the point of root elongation. Only a small percentage of the root hairs of a given plant become infected. The larger the number of the appropriate type of bacteria which are present in the soil, the more root hairs will become infected.

After legume plants are 3 to 4 weeks old and the seedling has developed its first true leaf, tiny nodules begin to be visible along

Figure 13.3 Proper inoculation of seed can result in more early-season forage, as shown by this test with arrowleaf clover in central Alabama.

the tap root. As plants become older, the number, size, and color of the nodules present will indicate the extent of N fixation. In older plants which have been properly inoculated, some nodules form on secondary roots, nodule size increases, and the interior of effective N-fixing nodules will be pink to dark red in color. A white or green color indicates the nodules are ineffective and not fixing N.

With time, legume nodules slough off, decompose, and the N becomes available to either legume or other plants growing in association with them. Many *Rhizobium* bacteria continue to live in the soil, assuming soil conditions are suitable. Amounts of N (lb/A/year) fixed by various legumes vary greatly depending on conditions, but good legume stands range from about 50 to 150 for annual clovers and vetches; 75 to 200 for birdsfoot trefoil, white clover, and red clover; and 150 to over 200 for alfalfa.

Where very little or no N fixation is occurring, the plants will indicate N deficiency by stunted growth and yellowing, especially the lower leaves. When inoculation has failed, it is sometimes possible to improve nodulation by broadcasting or drill-planting inoculated sand or by spraying an inoculum/water mix in the field, either of which should be done just prior to a rain or irrigation. Such an approach may or may not be effective, but is more likely to work if done soon after planting when seedlings are small.

Although legume seed inoculation is a simple process, it is highly important and deserves careful attention. The hot, dry conditions often encountered at planting time in the South can greatly reduce numbers of *Rhizobium* bacteria very quickly. Poor inoculation, and the resulting poor nodulation and N fixation, is a major reason why production of legumes is often difficult in the South. ■

Chapter 14

Forage Physiology

PHYSIOLOGY refers to the processes and activities associated with the functioning of a living organism. The growth pattern of forage grasses and legumes is influenced by physiological characteristics of the plant species and the ever-changing environmental conditions to which they are exposed. Physiological responses differ for annuals as compared to perennials and for cool season as compared to warm season species. Thus, an understanding of the physiology of forage plants can help in making better management decisions.

Light

Photosynthesis. Green plants have the capacity to capture energy in sunlight and use it to collect carbon dioxide from the air and water from the soil and convert them into products that can be consumed as food by animals. This process, photosynthesis, has a relatively low rate of efficiency, with light energy conversion of less than 3 percent for cool season grasses and about 5 to 6 percent for warm season grasses. These differences in efficiency are a result of different chemical pathways for the photosynthetic process. Cool season grasses such as tall fescue or ryegrass, as well as all legumes, fix energy into 3-carbon units and are called **C_3 plants**. Warm season grasses such as bermudagrass, pearl millet, or corn fix energy into 4-carbon units and are called **C_4 plants**.

The C_4 pathway in photosynthesis is more efficient for several reasons. Individual leaves of C_4 plants can utilize nearly full sunlight while C_3 plants become light-saturated at only 25 to 50 percent of full sunlight (**Figure 14.1**). The C_3 plants, unlike C_4 grasses, have **photorespiration** which may waste up to 40 percent of the energy captured in photosynthesis. Thus, C_4 grasses have the potential for much higher forage yields than C_3 plants.

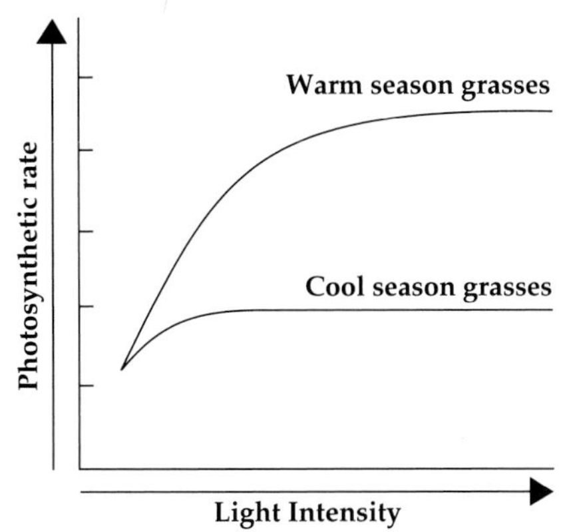

Figure 14.1. Photosynthetic potential of warm season (C_4) grass leaves is much higher than cool season (C_3) grasses.

Photosynthesis occurs in structures within plant leaves, called chloroplasts, in the presence of the pigment chlorophyll. Once carbohydrates have been formed, they can be rapidly translocated to various parts of the plant where they are used for metabolism, growth, or storage. Temperature, water availability, and other factors affect this translocation process.

The movement of materials within plants is made possible by specialized connective tissues which collectively are called the vascular system. The xylem transports water and minerals from the sites of absorption in the roots upward through the stem to the leaves. The phloem carries sugars and other compounds from the leaves where they are produced to other parts of the plants where they are used or stored.

Individual leaf photosynthesis. Nitrogen (N) deficiency and water stress greatly reduce the photosynthetic rate of leaves. Conversely, adequate fertilization and/or irrigation can enhance photosynthesis. Leaves that emerge in partial shade have a lower photosynthetic potential than leaves that emerge in full sunlight. The photosynthetic rate of individual leaves decreases with aging. The aging process occurs more rapidly under warm than cool conditions, with leaf life ranging from 30 to over 60 days. Thus, utilizing old leaves in a pasture to allow new young leaf growth can increase yield and quality.

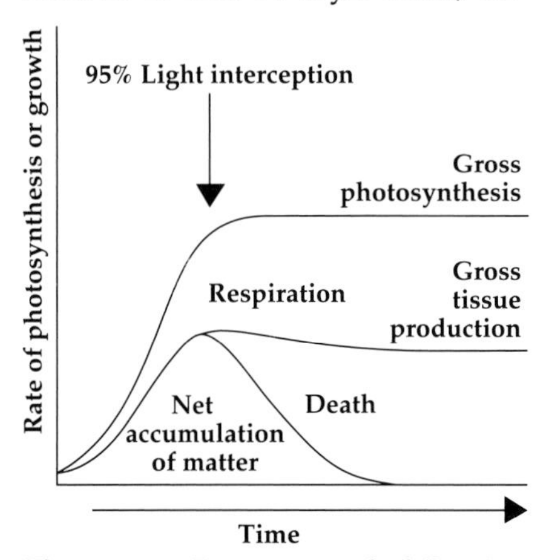

Figure 14.2. Forage growth following cutting, showing the balance between the rate of grass photosynthesis, loss in respiration, and death, and the rate of net accumulation of forage.
Adapted from: A.J. Parsons, E.L. Leafe, C. Collett, P.D. Penning, and J. Lewis. 1983. J. Appl. Ecol. 20:127-139.

Light relations in a pasture. The growth rate of a pasture or hayfield is related more to the total photosynthesis of the stand than to the photosynthesis rate of individual leaves. The reason for this is that leaves of forage grasses and legumes do not photosynthesize at their full potential in shade. This can be illustrated by the growth of a forage after cutting (**Figure 14.2**). The proportion of sunlight that is intercepted by individual leaves increases after cutting. Initially, all the leaves contribute to growth of the sward. However, as leaf numbers increase and shading of lower leaves increases, their photosynthetic contribution declines. This is a result of both shading and leaf aging. As leaf mortality increases, net accumulation of green matter decreases and gross tissue production remains about the same.

The arrangement of leaves within the canopy affects the amount of light intercepted and the photosynthetic potential of the sward. Horizontally displayed leaves as with white or crimson clovers absorb a greater amount of light than more vertical leaves such as those of tall fescue which allow light to penetrate deeper into the crop canopy. Plants with horizontal leaves need to be grazed or cut more frequently to avoid self-shading and rapid leaf death. Tall-growing legumes such as alfalfa allow more light deeper in the canopy, resulting in greater photosynthetic potential and higher yield. Warm season grasses such as corn, pearl millet or sorghum-sudangrass, with leaves set at an acute angle on tall stems, have a high photosynthetic potential and produce large forage yields over a short growing season.

An understanding of light relationships in a pasture can be helpful in devising management approaches to maximize productivity (see Chapter 25). Forage removal should be frequent enough to prevent the high leaf death losses occurring from shading and aging (**Figure 14.2**). However, after forage removal there should be sufficient green leaves left for photosynthesis to permit rapid growth. These objectives are often achieved under continuous stocking when adjustments are made at intervals to maintain short and desirable pasture height (see Chapter 25).

Regardless of the grazing method used, there is often a need to periodically mow or graze closely to stimulate new shoot production and regain the photosynthetic efficiency of a pasture. Light is needed to maintain a high population of active tillers (the leaf-producing unit) from buds at the basal portion of grass plants. Heavy shading in

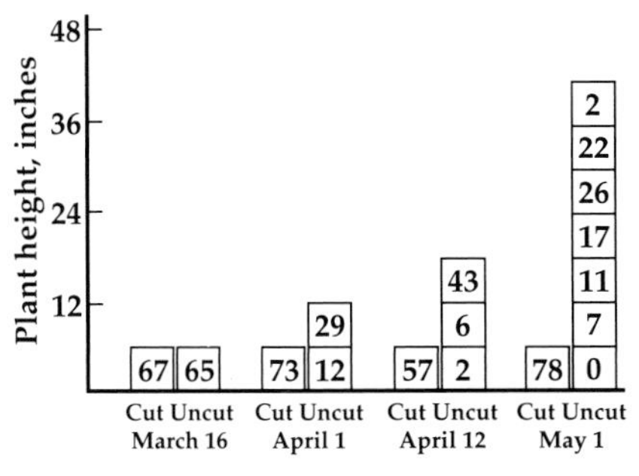

Figure 14.3. Distribution of buds per square foot (numbers in blocks) by 6-inch height intervals on Yuchi arrowleaf clover cut biweekly and uncut (central Alabama). Source: C.S. Hoveland, R.F. McCormick, and W.B. Anthony. 1972. Agron. J. 64:552-555.

undergrazed tall fescue pastures reduces tiller numbers for active autumn growth. Late-cutting of overmature tall fescue hay will result in low tiller populations and delay recovery growth for grazing.

Basal bud development in annual legumes such as arrowleaf clover is greatly affected by shading. Frequent defoliation will maintain a high population of new shoots and furnish grazing through late spring (**Figure 14.3**). When a hay cutting of arrowleaf clover is made in mid-April to late May in Zone A or the lower part of Zone B (**Figure 3.2**), little or no regrowth can be expected as there are few or no live buds remaining at the base of the plants. Light is essential for a high bud population in both grasses and annual legumes. Perennial legumes such as alfalfa and red clover maintain a relatively high number of buds on basal crowns and regenerate new shoots soon after each cutting. Birdsfoot trefoil has active branching at nodes above the ground so regrowth occurs there when light is adequate. With this legume, it is important to maintain several inches of leaves for rapid regrowth.

Competition for light is especially important where plant species are grown together. Successful competition for light depends on tolerance to low light levels and/or the potential to raise leaves above that of other plants. Although all pasture plants are sun-loving species, there are differences in tolerance to shading. Among the winter annual clovers, crimson and subterranean clovers are substantially more productive in shade than arrowleaf clover (**Table 14.1**). Tolerance to shading can be valuable for persistence of a species in an undergrazed pasture. But where frequent grazing

Table 14.1. **Effect of artificial shade on relative forage production of four annual clovers during winter and spring in the field at Athens, GA.**

Clover	Dry forage yield as percent of unshaded clover[1]		
	39% shade	64% shade	82% shade
Crimson	85	65	30
Subterranean	85	55	23
Berseem	75	48	20
Arrowleaf	63	40	18

[1]Unshaded clover yield for each species would be 100%.

Source: C.S. Hoveland and R.H. Brown, Univ. of Georgia. Unpublished data.

or cutting is done, this attribute will be of little value.

The proportion of different species in a mixture may shift substantially, depending on changes in light relations caused by cutting or grazing management. A closely-grazed tall fescue-ladino clover pasture is likely to have a high proportion of clover, but when undergrazed the clover will be shaded out and the pasture will become grass dominant. When a hybrid bermudagrass-alfalfa mixture is adequately fertilized with nutrients needed by alfalfa and cut for hay, alfalfa will dominate because it will shade out the bermudagrass.

Temperature

Species response to temperature. At optimum temperatures, the maximum growth rate of warm season grasses such as bermudagrass, pearl millet, and switchgrass is much greater than that of cool season grasses such as tall fescue and orchardgrass. The optimum temperature for warm season grasses is about 85 to 95°F, while the optimum for cool season grasses is about 60 to 80°F (**Figure 14.4**). Individual legume species differ some, with optimum temperatures being 65 to

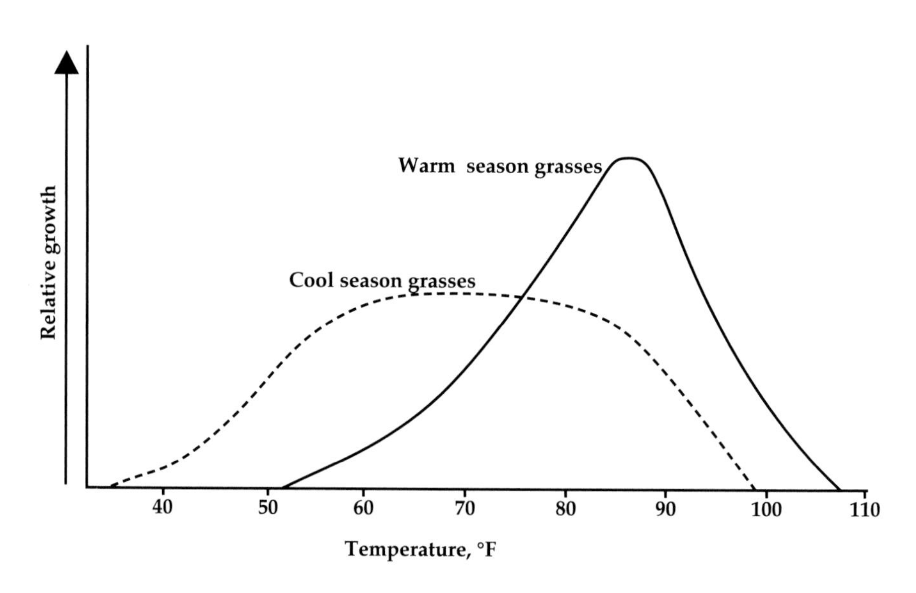

Figure 14.4. Relative growth rate of warm season grasses is greater than cool season grasses at high temperatures, but cool season grasses are productive over a much wider temperature range.

75°F for clovers, about 78°F for alfalfa, and 80 to 85°F for tropical legumes.

The temperature range over which forage is produced at near peak levels is much greater in cool season than in warm season grasses. Growth of warm season grasses falls off rapidly at temperatures below 70°F, with virtually no production at 50°F.

The rather narrow temperature range of warm season grasses means that the productive season gets progressively shorter as they are grown farther north where cool nights during spring and autumn limit growth. Based on temperature adaptation, warm season perennial grasses can be placed into three groups:

(a) Intolerant to cold and easily winterkilled, such as Coastcross-1 bermudagrass which has a long productive season of 7 to 8 months in the Gulf Coast area.

(b) Relatively cold-hardy, such as Coastal and Midland bermudagrass hybrids which have a shorter productive season the farther north they are grown.

(c) Cold-hardy, such as big bluestem, indiangrass, and switchgrass which can be grown as far north as Nebraska and Iowa.

Winter dormancy and winter productivity. In colder areas of the upper South winter dormancy among perennial forage varieties is important. Otherwise, they will begin growth during short warm periods in winter and consume carbohydrates needed for winter survival and early spring production. For this reason, alfalfa, red clover, orchardgrass, and tall fescue varieties grown in the upper South are winter-dormant types. Non-dormant types should not be grown in this region as they can be winterkilled or severely weakened by depletion of stored food reserves as a result of green-up during winter.

When winter-dormant varieties are grown in the lower South, they do not grow during the milder winter temperatures. Cool season forage plants of Mediterranean origin generally make more winter growth in the lower South than plants originating in colder climates. With less winter dormancy, they develop new leaves rapidly, with rapid leaf expansion during short periods of favorable temperature. This feature was exploited in the development of AU Triumph tall fescue, a winter-productive variety.

In central Georgia, AU Triumph yielded 128 percent more forage than winter-dormant Kentucky 31

Table 14.2. Autumn and late winter forage production of AU Triumph and Kentucky 31 tall fescue varieties in central and northern Georgia, 3-year average.

Location	Months	Dry forage yield	
		AU Triumph	Kentucky 31
		lb/A	
Central	Feb.-March	2,490	1,090
(Eatonton)	Nov.-Dec.	1,590	1,250
Mountain	March-April	1,980	1,790
(Blairsville)	Oct.-Nov.	1,230	930

Adapted from: C.S. Hoveland, M.W. Alison, Jr., G.V. Calvert, J.F. Newsome, J.W. Dobson, Jr., and C.D. Fisher. 1986. Georgia Agric. Exp. Stn. Res. Bull. 339.

tall fescue during February-March and 27 percent more during November-December (**Table 14.2**). When the two varieties were compared in the mountains of north Georgia (where temperatures are similar to southern Indiana), there were only small yield differences in response to the much colder and longer winter. When grown along the Gulf Coast, a non-dormant alfalfa such as Florida 77 remains green all winter and can be harvested from March through November. However, when this alfalfa variety is planted in the upper South, it can be expected to winterkill.

Warm season-cool season grass competition. Common bermudagrass and tall fescue can be found growing in mixtures from the lower to the upper South. When grown together, the competitive aspects of these grasses, one a C_4 and the other a C_3 species, are greatly affected by climate and other environmental variables. In the upper South, bermudagrass has a very short growing season. Temperatures are favorable for tall fescue growth during much of the growing season, therefore the tall fescue dominates a pasture or hayfield for most of the year.

Farther south, both grasses have restricted periods for favorable growth, and dominance by one or the other depends on weather conditions and management. During a wet year, tall fescue will dominate, while in a dry year bermudagrass will be most abundant. Nitrogen fertilization in autumn and late winter will favor tall fescue while fertilization in late spring and summer will favor bermudagrass. Close grazing in summer will further favor bermudagrass, especially at more southern locations. In the Coastal Plain, bermudagrass clearly has the competitive edge because of favorable temperatures over much of the year, sandy soils with high populations of nematodes that attack tall fescue roots, and often a long autumn drought to inhibit tall fescue growth.

In the mid-South where both warm and cool season grasses are widely adapted, it is often possible to favor one or the other by adjusting management. Therefore, proper management can result in a mixed warm and cool season grass pasture which provides grazing over much of the year. In addition to bermudagrass-tall fescue, other combinations of C_3 and C_4 plant species can be used, such as dallisgrass-white clover, annual ryegrass-arrowleaf clover overseeded on bermudagrass, and rye or ryegrass

Table 14.3 Water use efficiency of warm and cool season perennial forage plants in central Alabama, 2-year average.

Species	Water use efficiency, dry forage/inch of water
Warm season forages	
Coastal bermudagrass (C_4)	1,646
Common bermudagrass (C_4)	1,240
Pensacola bahiagrass (C_4)	1,194
Sericea lespedeza (C_3)	945
Cool season forages	
Tall fescue (C_3)	1,064
Orchardgrass (C_3)	1,060
Reed canarygrass (C_3)	1,005
Ladino clover (C_3)	480
Red clover (C_3)	436

Adapted from: B.D. Doss, O.L. Bennett, D.A. Ashley, and H.A. Weaver. 1962. Agron. J. 54:239-242.

O.L. Bennett and B.D. Doss. 1963. Agron. J. 55: 275-278.

overseeded on bahiagrass or bermudagrass.

Water

Water use. Water shortage is the major factor limiting plant growth, even when soil fertility and temperature are favorable. Although water is a major raw material in photosynthesis, less than 1 percent of the water taken up by the roots is used to produce food. The vast majority of plant water is transpired or lost by diffusion through the leaf stomata (tiny pores in leaves). Water escaping from leaves cools the plant and in its movement through the plant carries minerals, sugars, and amino acids. About 300 to 1,000 lb of water are used to produce one pound of dry matter.

Water use efficiency differs among forage species. The C_4 species (warm season grasses) are generally more efficient than C_3 plants in dry matter production per unit of water (**Table 14.3**). This is impressive, considering that the major production of C_4 species occurs during summer under high temperatures with high evapotranspiration rates, whereas the C_3 species (except a few species such as the lespedezas) make most of their growth during cooler periods of the year.

Water stress. Drought or moisture stress is a common problem in the South even though average annual rainfall is high (see Chapter 3). High summer temperatures cause high losses of water from plants and soil. High intensity, short-duration rainfall causes substantial runoff so that much of the rain does not enter the soil, particularly on clay soils such as in the Piedmont. Subsoil acidity adds to the problem by limiting deep soil penetration by roots of many species, especially legumes. Thus, with shallow root development much of the subsoil water is unavailable for plant growth. Forage species tolerant of subsoil acidity can utilize deeper soil water during periods of drought, furnishing higher yields. Thus, it is important to have deep-rooted forage species which maintain growth during periods of limited rainfall.

When soil moisture becomes limiting, many plant biological processes are affected. Cell division and enlargement are first affected by water stress. This stops tiller production and growth of water-stressed plants. When leaf growth and tiller development cease, photosynthesis continues at a reduced rate. This results in an increase in soluble carbohydrates, which can improve digestibility of drought-stressed forage. Minerals, especially potassium (K), accumulate in drought-stressed plants. Nitrogen is not metabolized into amino acids and proteins, resulting in accumulation in some plants of products such as nitrates and alkaloids which can be toxic to livestock consuming the forage (see Chapter 21).

As water stress develops, root growth may continue at a greater rate than foliage growth, thus allowing exploration of a greater volume of soil. This may result in the plants maintaining a reduced rate of photosynthesis over a considerable time as roots extract water from the soil. As water stress increases, photosynthesis will decrease, and eventually stop carbohydrate production. Leaf rolling in many grass species and folding of leaves in legume species is a common response to water stress. This reduces the heat load on the plant in the form of intercepted sunlight, and reduces transpiration water losses. In tall fescue, leaf rolling is

Table 14.4. Rooting depth and total roots in soil profile for various warm and cool season perennial forage plants in central Alabama, 2-year average.

Species	Total roots, lb/A	Maximum depth of roots, inches
Warm season forages		
Coastal bermudagrass (C_4)	4,762	78
Common bermudagrass (C_4)	3,950	79
Pensacola bahiagrass (C_4)	6,276	61
Sericea lespedeza (C_3)	3,085	50
Cool season forages		
Reed canarygrass (C_3)	12,768	48
Tall fescue (C_3)	9,011	48
Orchardgrass (C_3)	2,934	38
Ladino clover (C_3)	2,701	38
Red clover (C_3)	2,323	45

Adapted from: O.L. Bennett and B.D. Doss. 1960. Agron. J. 52:204-207.

B.D. Doss, D.A. Ashley, and O.L. Bennett. 1960. Agron. J. 52:569-572.

associated with endophyte infection and genetic factors. Research will further identify specific causes. If water stress is prolonged, premature leaf aging and death by desiccation occur.

Forage plants with roots that grow deep and have a high root volume with which to explore the soil should be most drought tolerant and productive during periods of low rainfall. Warm season perennial grasses are generally deeper rooted but do not necessarily have more total roots than cool season grasses (**Table 14.4**). For example, bermudagrass roots grow deeper in the soil but have a lower total weight of roots than bahiagrass, reed canarygrass, or tall fescue. Among the cool season species, reed canarygrass has a greater weight of roots but rooting depth is similar to tall fescue. However, root pruning by soil nematodes can drastically reduce the rooting depth of susceptible forage species. Tall fescue and orchardgrass are generally short-lived on sandy Coastal Plain soils because of root pruning by nematodes, increasing drought susceptibility.

Table 14.5. Relative yield (well-drained = 100) of clovers as affected by flooding over a 4-month period.

Clover	Flooding regime: 3 days in 10	6 days in 10	Continuously
Persian	100	82	85
Ladino	88	77	70
Ball	87	75	62
Intermediate white (S-1)	79	57	38
Berseem	76	60	52
Arrowleaf	71	46	36
Crimson	45	27	28

Adapted from: C.S. Hoveland and H.L. Webster. 1965. Agron. J. 57:3-4.

C.S. Hoveland and E.E. Mikkelsen. 1967. Agron. J. 59:307-308.

Flooding and poor drainage. Flooding and poor soil drainage reduce growth mainly by restricting oxygen supply to the roots. Consequently, forage species in a dormant state are much more tolerant of flooding than when actively growing. Oxygen requirements increase with temperature, thus plants often die under flooding or poor drainage. Also, some forage disease problems such as *Phytophthora* root rot in alfalfa increase.

Forage species differ greatly in their tolerance to poor drainage. Annual ryegrass, reed canarygrass, and dallisgrass are extremely tolerant of poor drainage, followed by tall fescue and johnsongrass. This contrasts with hybrid bermudagrass and orchardgrass which require good drainage for survival and production. Alfalfa is intolerant of poor drainage while clovers vary greatly (**Table 14.5**). Persian, ladino, berseem, and ball clovers are quite tolerant of, and productive on, wet soils. Arrowleaf and crimson clovers are intolerant of flooding and require good drainage for satisfactory production.

Carbohydrate Reserves

Photosynthesis supplies the energy needs of perennial forage plants for respiration and growth. Excess energy from photosynthesis is stored as carbohydrates in various plant parts. The main storage organs are roots in alfalfa, red clover, kudzu, birdsfoot trefoil, and sericea lespedeza; rhizomes in bermudagrass, bahiagrass, reed canarygrass, johnsongrass, and switchgrass; stolons in ladino clover; and stem bases in dallisgrass, tall fescue, orchardgrass, and big bluestem. These storage carbohydrates support respiration needs during winter or summer dormancy, allow regrowth after dormancy, promote flowering and seed formation, and contribute to cold and heat resistance of the plant. The levels of carbohydrates in storage organs are quite dynamic since many factors affect both their accumulation and use.

Cool season grasses such as tall fescue and orchardgrass depend on reserve carbohydrates for initiation of new leaf growth after a period of stress such as after a hay cutting, a period of very close grazing, a severe drought, or after winter in the upper South. In these species, continuous close grazing during summer reduces leaf area for photosynthesis and depletes carbohydrates so regrowth will be slow and stands weakened. This problem is more severe in the lower than in the upper South. Good management of these grasses should leave adequate leaf area for photosynthesis so carbohydrate levels are not depleted. Carbohydrates accumulate under conditions of low N, cool temperatures, and moisture stress. Carbohydrate depletion occurs when active growth is stimulated by N fertilization, adequate moisture, and higher temperatures. In contrast, low-growing, rhizomatous species such as Kentucky bluegrass can be grazed closely without damage.

Warm season grasses with erect growth habit such as johnsongrass, switchgrass, or big bluestem are vulnerable to damage by continuous close grazing and should be rotationally stocked to recharge carbohydrate storage organs. In contrast, bermudagrass and bahiagrass maintain large quantities of leaf area close to the ground and can be grazed closely without adverse effects.

Alfalfa, sericea lespedeza, and birdsfoot trefoil are especially sensitive to frequent close defoliation that depletes carbohydrates. Root carbohydrates in alfalfa have to be carefully managed in the northern USA and in the upper South where plants are dormant for several months in winter. Respiration of roots during this dormant period utilizes energy in northern areas, so it is important to not harvest alfalfa for 4 weeks before a killing frost to allow adequate carbohydrate accumulation. Otherwise, plants will be weakened, winter injury will be increased, and forage yields the following spring will be decreased.

In the lower South, alfalfa root carbohydrates are depleted after each cutting, but they are recharged (**Figure 14.5**). Since in this region alfalfa remains green and maintains a low rate of photosynthesis during autumn, it is not necessary to leave it uncut prior to frost. Thus, alfalfa can be cut for hay or silage every 30 to 40 days throughout the growing season in the lower South.

When alfalfa, sericea lespedeza, or birdsfoot trefoil are closely and continuously grazed, carbohydrates are rapidly depleted and both yields and stands will decline. Rotational stocking is imperative to maintain productivity of these species (see Chapter 25). Grazing-tolerant alfalfa varieties maintain more leaves near the soil surface, higher root carbohydrate levels, and can better tolerate continuous stocking. Ladino clover, a prostrate species with stolons, can be grazed closely with no stand loss. Winter annual clovers do not accumulate carbohydrates for regrowth but depend on residual leaf area and rapid leaf regeneration from active buds.

Vernalization and Photoperiod

An initial period of spring vegetative growth is normally followed by a reproductive period when flowering and seed production occur. Most cool season perennial grasses must be exposed to winter conditions (low temperatures and long nights) to initiate flower buds. Temperatures between 32 and 50°F (vernalization) and about a 16-hour night initiate flower buds. Later in spring, short nights and mild temperatures cause inducted buds to grow to flowering stems. Warm

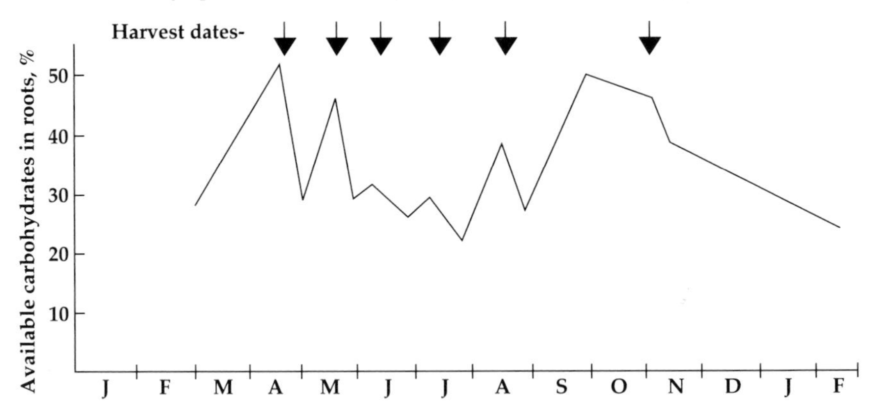

Figure 14.5. Seasonal trend in percent available carbohydrates in alfalfa harvested for hay at Athens, GA.

Source: L.G. Brown, C.S. Hoveland, and K.J. Karnok. 1990. Agron. J. 82:267-273.

season grasses such as dallisgrass and bahiagrass flower in early summer in response to short nights and warm temperatures. Some warm season grasses are extremely sensitive to short variations in night length which affect flowering.

Plants can be grouped into three categories based on their flowering response to daylength: (a) Plants that flower during spring when the daylength is increasing are referred to as "long-day" (short night) plants. Cool season grasses and legumes fit this category. (b) Plants that flower only in summer or early autumn when daylength is decreasing are "short-day" (long night) plants. Examples include annual lespedeza, sericea lespedeza, and kudzu. (c) Other plants such as alfalfa may flower throughout much of the growing season and are referred to as "day-neutral". Though of little practical importance, it is the length of the uninterrupted dark, rather than light, which is responsible for the flowering response in long-day and short-day plants.

Plant Hormones

There are numerous types of plant hormones including auxins, gibberellins, cytokinins, and ethylene. Each of these compounds has important influences on specific aspects of plant growth. Among other processes, plant hormones are responsible for the downward growth of roots and the upward growth of stems in response to gravity (geotropism), the leaning of plants toward a light source (phototropism), stem elongation of plants deprived of adequate light (etiolation), and rapid cell division and elongation at growing points.

Seeds

Seeds are the key organs of propagation, preserving the genetic characteristics of the plant and dispersing the genetic diversity generated during reproduction. A seed is a young plant in a resting stage but if stored too long, especially under poor storage conditions, seed will eventually die. Satisfactory seed germination requires seed capable of germinating, favorable temperature (which varies with forage species), adequate moisture (97% relative humidity at 66°F), and sufficient oxygen (no prolonged flooding). Rapid uptake of water during the first few hours of germination initiates enzyme mobilization of stored food reserves for seedling development. The rate of seedling development is then determined by the temperature and oxygen supply. Thus, for successful seed germination it is important to plant high-quality seed at the proper time of year at the correct depth in close contact with the soil.

Summary

Forage physiology deals with how plants respond to light, temperature, water, and nutrients and also the growth processes of different plants, including forage grass and legume species. Growth of a plant canopy depends upon the area and arrangement of leaves for light interception and photosynthetic potential. It also depends on maintaining a dynamic population of tillers for replacement of dying or harvested leaves. Understanding the interactions of these variables and the responses can assist in developing sound grazing or cutting systems that maintain productivity and forage quality over as much of the growing season as possible. ■

Chapter 15

Weeds in Forage Crops

IF A GROUP of forage producers selected at random from throughout the South were asked to list their main forage production problems, weeds would almost certainly be high on most of their lists. A short drive through almost any part of the region will quickly provide evidence that better weed control is needed in many pastures and hayfields (**Appendix A.34**).

One of the best definitions of the term "weed" is, "A plant growing in a setting in which it is considered undesirable." If a plant is undesirable in a forage crop situation, the clear (and correct) implication is that it is in some way inferior to preferred forage species.

Why Control Weeds?

There are several good reasons for controlling weeds in forage crops. They usually reduce yield and often lower forage quality. They also compete with desirable forage plants for moisture, nutrients, sunlight, and space. In addition, some weeds are toxic, while others are unpalatable or injurious to grazing livestock because of thorns, spines, or certain distasteful chemical constituents.

However, weeds are not always totally bad in a forage situation. Weeds help prevent soil erosion in areas where a good forage stand has not been obtained. In addition, weeds often have some nutritive value. **Tables 15.1 and 15.2** provide evidence that many weeds commonly found in pastures in the South have surprisingly high nutritive value for livestock.

Nonetheless, these data do not account for the fact that some weeds

Figure 15.1. Weeds compete with desirable forage plants for moisture, nutrients, light, and space.

Table 15.1. Crude protein and in-vitro dry matter digestibility (IVDMD) of selected cool season weeds and forages at three stages of maturity.[1]

	Vegetative		Flower/boot		Fruit/head	
Weeds	Crude protein (%)	IVDMD (%)	Crude protein (%)	IVDMD (%)	Crude protein (%)	IVDMD (%)
Carolina geranium	19	78	14	70	11	68
Cutleaf evening primrose	20	72	14	69	11	52
Virginia pepperweed	32	86	26	72	17	63
Curly dock	30	73	19	54	16	51
Virginia wildrye	23	80	19	74	7	60
Wild oats	23	75	—	—	—	—
Cheat	23	81	18	69	14	61
Little barley	24	82	18	78	14	62
Forages						
Rye	28	79	24	81	13	70
Tall fescue	22	78	17	73	13	67
Ladino clover	27	81	22	85	23	83
Hairy vetch	30	80	29	77	26	77

[1]Adapted from: C.S Hoveland, G.A. Buchanan, S.C. Bosworth, and I.J. Bailey, 1986. Alabama Agric. Exp. Stn. Bull. 577.

Table 15.2. Crude protein and in-vitro dry matter digestibility (IVDMD) of selected warm season weeds and forages at three stages of maturity.[1]

	Vegetative		Flower/boot		Fruit/head	
Weeds	Crude protein (%)	IVDMD (%)	Crude protein (%)	IVDMD (%)	Crude protein (%)	IVDMD (%)
Sicklepod	22	84	14	76	17	71
Coffee senna	17	81	22	75	15	67
Hemp sesbania	31	70	14	66	11	52
Tall morningglory	20	82	—	—	14	76
Ivyleaf morningglory	19	80	—	—	11	78
Cypressvine morningglory	21	80	—	—	13	77
Florida beggarweed	22	74	17	65	13	56
Prickly sida	17	80	18	70	12	56
Common purslane	—	—	19	80	—	—
Bur gherkin	—	—	17	75	14	79
Redroot pigweed	24	73	17	71	11	64
Jimsonweed	25	72	21	66	17	56
Fall panicum	19	72	9	63	7	54
Texas panicum	16	74	11	62	8	52
Yellow foxtail	18	73	12	66	14	57
Crabgrass	14	79	8	72	6	63
Crowfootgrass	16	67	8	54	9	43
Forages						
Pearlmillet	17	59	6	60	8	60
Bermudagrass	16	58	7	51	8	43

[1]Adapted from: C.S. Hoveland, G.A. Buchanan, S.C. Bosworth, and I.J. Bailey, 1986. Alabama Agric. Exp. Stn. Bull. 577.

are unpalatable, produce low yields, or have a short productive season. If there were no disadvantages associated with these plants, they would not be considered weeds and would be recommended for forage plantings.

Origin of Weeds in Forage Crops

Weed problems in pastures do not just "suddenly develop." Weeds originate primarily from seed, so the question actually is, "What was the source of the seed?" Usually when a weed problem develops, there have been low weed populations which have gone largely unnoticed for several years. During this time, the weeds were producing seed which could lie dormant until a situation occurred which favored germination and establishment.

Seed of some weeds can lie dormant in the soil for many years. In fact, the seed may have been in the soil in a pasture for many years before the present producer used the area. This is particularly true for weed populations which develop after perennial grass pastures are plowed or tilled for the first time in many years. Thus, weeds may appear which the producer has never seen in the area. Showy crotalaria (a poisonous plant) is a good example of a weed capable of surprising landowners in this manner.

Most livestock producers know that weed seed are frequently present as a contaminant in crop seed they purchase. Using certified crop seed will reduce weed contaminants and is highly recommended (see Chapter 10). "Cheap" seed are no bargain. In addition, weed seed can be brought onto a farm in hay or crop residues which may be fed to livestock, such as soybean tailings and crop straw or stubble.

Weed seed are also transported by natural means such as wind, and/or water, as well as birds and wild animals. Farm livestock also disseminate seed, either by carrying them on their hair coats or hooves, or by eating them and later depositing them in dung. Because of these varied means of movement, a severe weed problem which develops on one farm often spreads to neighboring farms.

Most weeds are prolific seed producers and have what might be called a "shotgun" approach to propagation. The seed are dispersed at random and some find suitable conditions for germination and establishment. Most pastures in the South contain plenty of weed seed which, under suitable conditions, can become established.

Soil type and site have a great influence on the likelihood of weeds becoming a problem. For example, sedges and rushes are best adapted to moist areas, while prickly pear and yucca (bear grass) are found in relatively droughty areas. In many cases, a slight edge in adaptation over pasture species or other weeds allows a weed to dominate in a particular area.

Stocking rate and forage utilization also have a profound influence on the species composition of a pasture. For example, many clovers are much more likely to persist in an area in which frequent clipping or close grazing keeps grasses at a low height. Johnsongrass, which is a serious weed in row crops but an important forage crop in some areas of the South, will not persist under close grazing or frequent cutting.

The population density of many weeds is greatly affected by grazing management. For example, the palatability of broomsedge is never very good, but livestock will consume young, tender broomsedge growth. However, the plant becomes **extremely** unpalatable as it becomes older, causing livestock to avoid it under almost any circumstance.

This characteristic allows broomsedge to become established in situations where pastures are undergrazed in spring and overgrazed in mid to late summer. With undergrazing in spring, young broomsedge plants are allowed to become mature. If pasture growth becomes short in summer, as periodically occurs in most of the South, the livestock refuse the broomsedge and heavily graze improved species, thus reducing their competitiveness.

Some weeds, especially certain grass weeds, are favored under intensive grazing. Lower stocking rates are more likely to favor broadleaf weeds.

Weed Control Approaches

Liming, clipping, grazing, and use of herbicides can discourage or kill weeds while encouraging growth of desirable forage species. It is important to provide the proper soil pH and fertility levels for desired forage species, both during establishment and later in maintenance. Weeds usually cannot compete well with dense stands of improved forage species which have been adequately limed, fertilized, and properly grazed. Maintaining adequate soil fertility is usually effective in preventing weed encroachment once a good forage stand has been obtained, but is less effective in eliminating weeds once they have become established.

Periodic clipping of pastures also provides some weed control benefits if done at the proper time. Clipping kills many annual broadleaf weeds if they are cut off below their growing points. In addition, clipping can prevent or reduce weed seed production which helps reduce future weed problems.

The frequent, non-selective clipping exercised in hayfields, together with the generally higher fertility levels which stimulate forage crops to be competitive, explain why there are fewer weed problems in hayfields than in pastures. Clipping pastures also has the advantage of stimulating new forage growth which is of higher quality than older, more mature forage.

Goats can be useful in control of some weeds because they graze some troublesome species which are unacceptable to cattle. For example, goats relish woody and/or spiny weed species such as persimmon, briars, and spiny amaranth. In addition, curly dock is neither spiny nor woody, but is unpalatable to cattle while being a favorite of goats.

In many cases, weeds in forages can be controlled with herbicides. The wise use of herbicides according to label directions is safe to humans, animals, and crops, and often can effectively and economically increase forage productivity. However, other management practices must be employed to maintain an improved forage situation. The dynamic nature of herbicide recommendations precludes providing specific herbicide information in this book, but such information is readily available from agricultural Extension agents and other sources.

Herbicide Application Tips

Keeping a few basic concepts in mind when making plans to spray a pasture for weed control can greatly increase the likelihood of success:

(1) Match the herbicide and rate of application to the weeds and pasture crops which are present. Although the number of herbicides labeled for use on pastures is limited, producers should become familiar with their herbicide options for each forage crop they grow.

(2) Herbicides must be applied at the correct time to be effective. Weeds are easiest to kill when they are small, preferably no

Figure 15.2 Herbicides can often provide economical weed control in forage crops.

more than 4 to 6 inches in height. If weeds get taller than about knee-high, it is desirable to clip the pasture and spray the young regrowth.

(3) Weeds should be sprayed when they are young and actively growing. A severe drought period during which weeds are making little growth is not a good time to apply a herbicide, but spraying a few days after a drought has ended is an excellent practice. The extent to which weeds are actively growing when sprayed often determines the degree of success.

(4) Most pasture herbicides are most effective at 60°F or above, and it is helpful to have several warm days in succession before and after a herbicide application is made. Windy weather will prevent good coverage, and rain within a few hours of spraying will reduce effectiveness.

(5) It is necessary to get good herbicide coverage in order to obtain adequate weed control. At least 15 to 20 gallons of water per acre should be used with most herbicides, and spray pressure should be at least 30 to 40 p.s.i. Some herbicides require the addition of a surfactant or spray adjuvant. Using liquid fertilizer solutions as the spray carrier may cause forage injury in some cases. The herbicide label should be consulted regarding these points. Any grazing, haying, or other restrictions on the label should be strictly observed.

(6) Pasture herbicides should always be used in a safe manner. Crops in the vicinity of the area to be sprayed may be damaged by spray or vapor drift. Cotton and tomatoes, for example, are extremely sensitive to phenoxy herbicides. In some cases an interplanted forage crop such

as clover or winter annuals may be at risk if a herbicide is applied. Also, without exception, safety of the operator must be a primary consideration.

(7) **Herbicide labels should always be read carefully prior to use of the material, and the directions followed explicitly.** While most pasture herbicides are relatively non-toxic, one should always be careful when handling concentrated pesticides or spray mixtures.

Summary

Weeds in forage crops are undesirable, but the severity of the problem depends on the type and population of weeds present.

In general, managing forage species in ways which allow them to be competitive provides the cheapest, most effective, and longest-lasting weed control. Taking care to prevent the introduction of weed seed onto a farm is another important way of reducing the likelihood of having weed problems. Periodic clipping and/or herbicide applications are other options which should be exercised in some situations. A forage producer may or may not be responsible for weed problems developing, but **is** responsible for allowing them to remain. ■

Chapter 16

Forage Quality

FORAGE QUALITY is best defined in terms of animal performance, such as daily gain, milk production, wool production, or reproduction. Many attempts have been made to define forage quality in other ways, including leafiness, fineness of stem, color, smell, protein content, fiber content, and amount of lignin. Chemical composition of forages is a useful, but imperfect, measure of forage quality. It is the animal rather than the human that ultimately determines forage quality.

Palatability

Palatability refers to the preference which livestock show for one forage over another. Generally, high quality forages are highly palatable, and vice-versa. But not always. Animal selection of one forage species over another depends on smell, touch, and taste. Palatability may be affected by texture, aroma, succulence, hairiness, leafiness, fertilization, dung pats, urine patches, sugar content, or some component that causes a forage to be sweet, bitter, sour, or salty.

When animals are restricted to a less palatable species, animal performance may be quite satisfactory if forage nutritive quality is high and intake is not reduced. For example, grazing animals will select oats in preference to rye even though they will perform well when restricted to rye pasture (**Figure 16.1**). Annual ryegrass is even more palatable than oats. Thus, palatability alone can be a misleading indicator of forage quality. In

Figure 16.1. In cafeteria-style plots, horses consistently selected and grazed oats (being grazed) or annual ryegrass as compared to rye (right), wheat, or triticale. Athens, GA.

addition, palatability of a forage species may change during the growing season.

Forage Digestion by Animals

Ruminant animals (i.e., cattle, sheep, goats, deer, camels, buffalo, and llamas) have an enlarged 4-compartmented stomach which permits them to digest large amounts of forage with a high fiber content. This unique and valuable characteristic enables these animals to utilize materials which cannot be efficiently utilized by non-ruminants. Forage consumed by ruminants moves from the esophagus into the rumen and reticulum, then is regurgitated and chewed before being digested by rumen microflora (**Figure 16.2**).

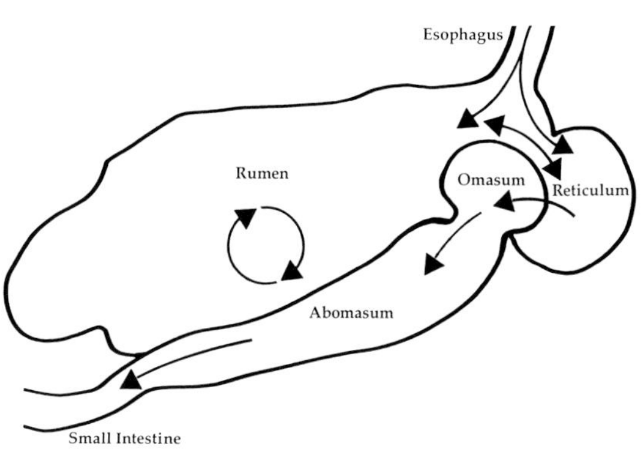

Figure 16.2. Representative movement of forage through the ruminant animal digestive tract.

Volatile fatty acids from the digestion process are absorbed through the rumen wall and utilized by the animal. The microbes and undigested material pass into the omasum where water is absorbed. Then the microbes are degraded in the abomasum (true stomach) and the products absorbed in the small intestine for use by the animal. Starch and sugars are absorbed in the small intestine. **Digestibility** (the proportion of the forage consumed in passage through the alimentary tract) varies greatly, depending on the type of material consumed by the animal. Immature leafy grasses may be 80 to 90 percent digestible, while mature stemmy material is often below 50 percent.

Certain non-ruminant animals such as horses have a functional caecum and colon where microbes digest fiber. However, the caecum (the fermentation vat) and colon follow the small intestine, and the total stomach capacity is much smaller than for ruminants; hence microbial activity is low in the horse. The horse breaks down only about 30 percent of the fiber in the forage, whereas the ruminant utilizes 60 to 70 percent. Horses cannot digest as much low-quality roughage as ruminants, thus higher quality forage must be fed for good performance.

Forage Composition

Leafy, growing forage plants contain 70 to 90 percent water. Thus, it is preferable to express forage yields and chemical analyses on a dry matter basis. The constituents of forages can be divided into two main categories: 1) those present as cell contents or the non-structural part of the plant tissue (protein, sugar, and starch); and 2) those which make up the structural components of the cell wall (cellulose, hemicellulose, and lignin).

Cell contents are almost completely digested, while the digestibility of the cell walls varies due to many factors, including species, age, and temperature. In addition, minerals, vitamins, and anti-quality factors such as tannins, nitrates, alkaloids, cyanoglycosides, estrogens, or mycotoxins influence animal performance, depending on the plant species and environmental conditions.

Protein

Digestible proteins are located in the cell cytoplasm (cell contents) and relatively indigestible proteins are found in the chloroplasts (structures located within the cytoplasm) where they make up about 40 percent of the total protein. Cell walls tend to retain large protein molecules, often making them unavailable to the animal during digestion. Chemical analyses for protein are expressed as "crude protein", indicating some of these proteins may be unavailable for animal utilization.

The portion of the crude protein which is unavailable varies with crop species, stage of maturity at harvest, and other factors. In plants such as bermudagrass and bahiagrass, which have highly lignified cell walls, much of the crude protein may be unavailable to the animal. However, protein quality is not of great importance for beef cows on pasture since all amino acids can be synthesized by rumen microorganisms.

Young, leafy grasses and legumes are normally high in protein and usually meet the protein needs of grazing animals. Perennial cool season grasses generally contain less protein than legumes, but have adequate amounts for most types and classes of animals. Winter annual pastures such as small grains and annual ryegrass contain high levels of crude protein, yet growing stocker cattle gain faster when supplemented with rumen bypass protein feeds such as fish meal or certain grain components. The reason is that a large portion of the digestible nitrogen (N) in the forage is soluble and is quickly converted to ammonia in the rumen and lost.

Well-fertilized warm season grasses in pasture, or hay cut at early stages of maturity, generally contain adequate protein for beef cows. However, insufficient protein can be a problem in warm season grasses with inadequate N fertilization, but especially on frosted pasture during winter or in hay cut at advanced stages of maturity.

Digestible Energy

The word "carbohydrate" refers to chemical compounds which are composed of carbon, hydrogen, and oxygen. Carbohydrates are extremely important nutritionally because they are the main source of energy in food. Non-structural carbohydrates in forages are readily digested by animals of all types. However, ruminant animals have the unique ability to digest some of the structural carbohydrates located in the cell walls of plants.

Cellulose and hemicellulose are structural carbohydrates which may be digested by bacterial action in the rumen, but this is sharply reduced as lignin content increases. Lignin, indigestible in ruminant animals, is low in legumes such as white clover and birdsfoot trefoil or immature small grains and ryegrass. Alfalfa, a high-quality legume at immature growth stages, declines rapidly in digestibility as stem lignin content increases with maturity. Warm season perennial grasses such as bermudagrass and switchgrass develop lignin rapidly as they mature, thus reducing their digestibility. Highly lignified forages remain in the rumen for long periods of time because of their slow rate of digestion, thus decreasing dry matter intake. Animal performance is markedly reduced by the low digestibility of lignified forages due to a reduction in forage consumption.

Digestible energy is generally the most limiting nutritive factor affecting forage intake and animal performance. Animal production falls off rapidly and weight loss may result when highly lignified overly-mature warm season perennial grasses are consumed. Forage utilization and animal performance are dependent on digestibility and voluntary intake of the forage. Generally, there is a high correlation between digestibility and forage intake and between intake and animal performance.

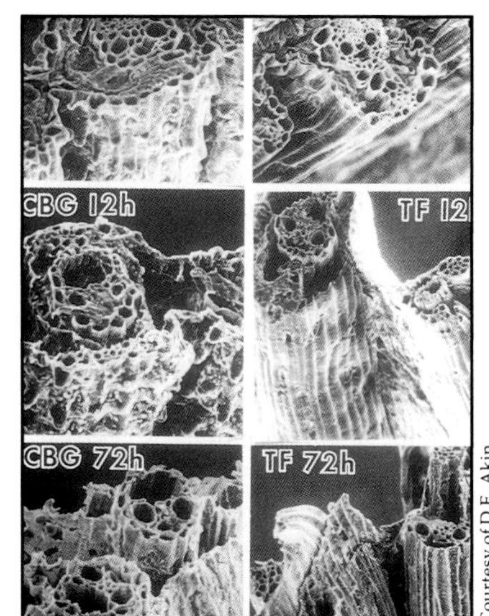

Figure 16.3. Coastal bermudagrass (CBG) forage (left) has slower digestibility than Ky 31 tall fescue (TF) forage (right). Undigested portions of grass are lignified structural parts remaining after 0, 12, and 72 hours in rumen fluid.
Source: D.E. Akin, USDA Russell Research Center, Athens, GA.

Factors Affecting Forage Quality

1. Plant species.

Cool season grasses are generally more digestible than warm season grasses. For instance, tall fescue forage is more digestible than Coastal bermudagrass (**Figure 16.3**). Furthermore, cool season annual species (ryegrass, oats, wheat, and rye) are more digestible than cool season perennial grasses (orchardgrass, tall fescue, or Kentucky bluegrass) at the same stage of maturity. There are also differences among species having similar seasonal distribution of growth, as evidenced by warm season grasses such as dallisgrass and Johnsongrass, which are more digestible than bahiagrass, bermudagrass, or carpetgrass at the same stage of maturity.

Breeding work with bermudagrass has resulted in significant improvement in digestibility. Coastcross-1 is approximately 12 percent more digestible than Coastal bermudagrass, resulting in a 30 percent higher average daily gain of beef steers. Tifton 78 and Tifton 85, more cold hardy than Coastcross-1, are also more digestible than Coastal.

Legumes such as alfalfa, clovers, and birdsfoot trefoil are generally of high quality, and digestibility falls less rapidly with maturity than is the case for warm season perennial grasses. Legumes also generally contain higher levels of protein than grasses.

Plants can also contain unique compounds which have an adverse effect on forage quality. For example, tannins in some sericea lespedeza varieties reduce digestibility, intake, and animal performance. Fortunately, plant breeders have developed low-tannin sericea lespedeza varieties which result in improved animal performance.

2. Climate.

Digestibility of warm season and cool season perennial species is highest in spring, falling to a low level in mid to late summer, and increasing in autumn. Even alfalfa under rotational stocking declines in digestibility and crude protein during hot summer weather (**Table 16.1**).

Drought stress has little effect on forage quality as long as the plants remain alive. Moderate stress may actually increase forage digestibility.

Table 16.1. Seasonal change in in-vitro dry matter digestibility (IVDMD) and crude protein (CP) content of Apollo alfalfa pasture in central Georgia, 2-year average. Pastures were sampled at the beginning of each 2-week grazing period followed by 4 weeks of rest in a rotational system.

Month	IVDMD	CP
	------------ %	------------
April	69	25
May	65	22
June	61	18
July	57	16

Adapted from: C.S. Hoveland, N.S. Hill, R.S. Lowrey, Jr., S.L. Fales, M.E. McCormick, and A.E. Smith. 1988. J. Prod. Agric. 1:343-346.

Animal gains on drought-stressed forage are often above average as long as adequate quantities are available. Excess rainfall may increase the moisture content of forage, having no effect on digestibility but often reducing dry matter intake. However, excess rainfall may reduce protein in grasses as a result of leaching N from the soil.

High temperature lowers digestibility by increasing lignification. Tannin content of sericea lespedeza also increases with high temperature, reducing palatability and digestibility when grazed (**Table 16.2**). However, tannin levels decline in sericea lespedeza hay.

3. Stage of maturity.

Maturity has a greater effect on nutritive value than does any other factor. As plants mature and cell walls become more lignified, they constitute a larger proportion of the cell and result in an overall decrease in digestibility and crude protein content. This decline in quality with maturity is greater and more rapid in warm season than cool season grasses.

Table 16.2. Seasonal changes in forage tannin concentration of high- and low-tannin sericea varieties at Athens, GA, 3-year average.

Variety	Tannin content		
	June	August	October
	----------	%	----------
Appalow (high-tannin)	7.9	12.8	10.5
Serala (high-tannin)	8.0	11.6	9.5
AU Donnelly (low-tannin)	4.3	5.3	5.3
AU Lotan (low-tannin)	3.1	4.9	4.6

Adapted from: C.S. Hoveland, W.R. Windham, D.L. Boggs, R.G. Durham, G.V. Calvert, J.F. Newsome, J.W. Dobson, Jr., and M. Owsley. 1990. Georgia Agric. Exp. Stn. Res. Bull. 393.

As bermudagrass becomes more mature, forage yield increases but digestibility and crude protein decline substantially (**Figure 16.4**). At 7 weeks of age, forage quality can be very low, with digestibility below 50 percent and crude protein below 8 percent.

Cool season perennial grasses such as tall fescue also decline in nutritive quality with advancing maturity in spring (**Figure 16.5**). Crude protein remains high until

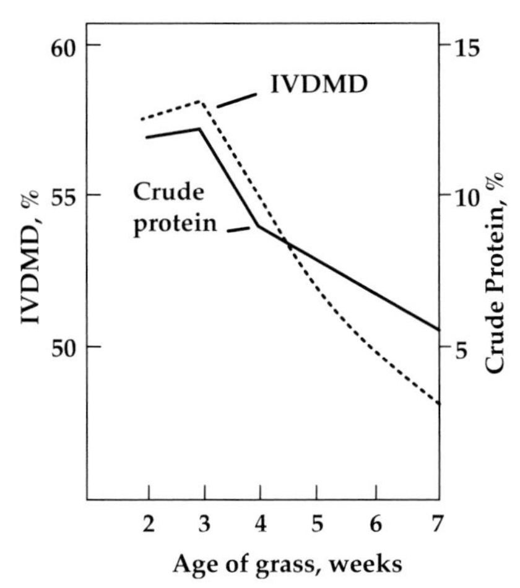

Figure 16.4. In-vitro dry matter digestibility (IVDMD), and crude protein content of Coastal bermudagrass as affected by plant maturity in south Georgia.
Source: W.G. Monson, USDA, Coastal Plain Exp. Stn., Tifton, GA. Unpublished data.

seed dough stage but digestibility falls drastically after late boot. The NDF levels are high at seed dough stage, predicting reduced animal intake. Well-fertilized aftermath tall fescue harvested in summer can have crude protein and IVDMD values of 15 and 56 percent, respectively. Autumn-harvested tall fescue is of high quality and can have crude protein values of 15 to 17 percent and IVDMD levels of 63 to 65 percent.

Immature grass may contain 65 percent soluble cell contents, and 35 percent cell walls. However, at a mature stage the same grass species may contain only 40 percent cell contents but 60 percent cell walls of which 7 percent may be insoluble lignin. Crude protein content may decline from 30 percent at an immature stage to only 5 or 6 percent when mature.

Legume species likewise differ in digestibility as they mature. Ladino

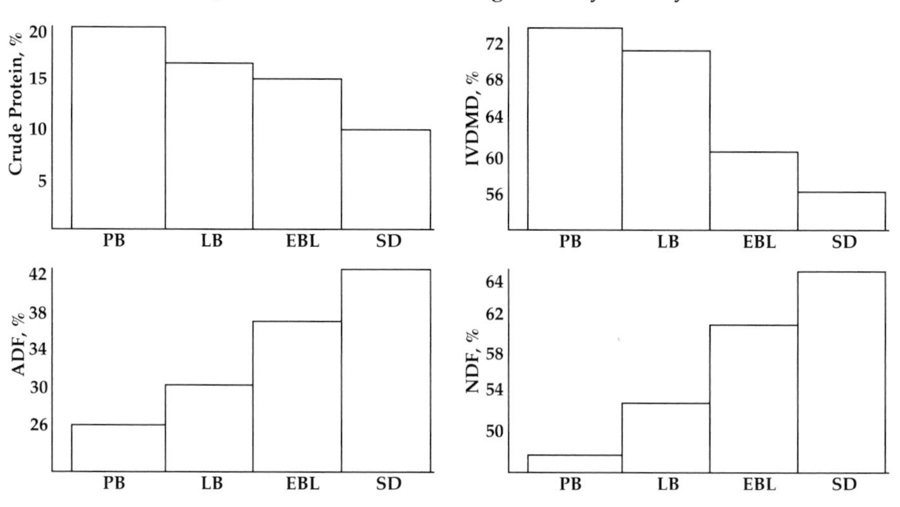

Figure 16.5 Changes in Kentucky 31 tall fescue crude protein, in-vitro dry matter digestibility (IVDMD), acid detergent fiber (ADF), and neutral detergent fiber (NDF) as affected by stage of maturity at first harvest: PB = pre-boot, LB = late boot, EBL = early bloom, and SD = seed dough. Athens, GA. 2-year average.
Source: C.S. Hoveland and N.S. Hill, University of Georgia, unpublished data.

clover, lacking true stems, maintains its high quality over time. Birdsfoot trefoil stems remain highly digestible as the plants mature. Alfalfa stems become lignified and digestibility drops sharply after early bloom (**Table 16.3**). Among the annual clovers, arrowleaf stems remain more digestible than crimson clover as they mature.

4. Nitrogen fertilization.

Up to a point, application of N fertilizer to grasses increases crude protein content, assuming other nutrient elements are not highly deficient, thus greatly limiting growth. However, N fertilization generally has little effect on digestibility of young leaf material. Harvesting, curing, and storage of cut forage can greatly affect digestibility (see Chapters 18 and 19).

Table 16.3 Relationship of the stage of alfalfa maturity at harvest to total digestible nutrients (TDN), crude protein (CP), and acid detergent fiber (ADF).

Maturity	TDN	CP	ADF
	----------	%	-----------
Pre-bud	65	21.7	28
Bud	62	19.9	31
1/10 bloom	58	17.2	34
1/2 bloom	56	16.0	38
Full bloom	54	15.0	40
Mature	52	13.6	42

From: Nutrient requirements of dairy cattle. 1978. Nat. Acad. Sci. Publ. 1349.

Methods of Measuring Forage Quality

Laboratory evaluation of forages is an attempt to predict the extent and rate of biological degradation in the ruminant animal. The objective is to be able to formulate diets from compositional information that will result in predictable animal response.

1. Proximate analysis. This laboratory technique is the oldest and has been the most commonly used form of forage evaluation. The expression of total digestible nutrients (TDN) attempts to equate feeds on an energy basis. The major problem with this method is that no component analyzed represents a single chemical substance. Further, the data are of limited value since the extent to which cellulose, hemicellulose, and total cell wall contents are digestible differs among plant species and varieties and also varies with plant maturity.

2. Neutral-detergent fiber (NDF) and acid-detergent fiber (ADF). These methods chemically distinguish the readily available and soluble cell contents from the less digestible portion of cell walls. Neutral detergent fiber consists of cell walls while ADF represents lignified cellulose. Neutral detergent fiber values are negatively correlated with forage voluntary intake by the animal while ADF (NDF less the hemicellulose component) is negatively correlated to digestibility (see Appendix A.8).

3. In-vitro fermentation. In this procedure, the forage is digested in rumen fluid followed by treatment with acid pepsin to simulate the digestive process in the ruminant animal. Results from this procedure give a good approximation of digestibility, causing this to be a widely used method for comparing digestibility.

4. Near infrared reflectance spectroscopy (NIRS). This rapid, reliable instrument technique is based on the amount of light reflected from, or transmitted through, a finely ground forage sample. The

Table 16.4. **Total digestible nutrients (TDN), in-vitro dry matter digestibility (IVDMD), crude protein (CP), neutral detergent fiber (NDF), and acid detergent fiber (ADF) of four perennial grasses.**[1]

Grass	Age at harvest	TDN	IVDMD	CP	NDF	ADF
	weeks	----------------		%		----------------
Coastal bermudagrass	4	55	62	13	58	29
Coastal bermudagrass	8	49	47	8	65	40
Tall fescue	4	66	63	14	58	31
Orchardgrass	4	66	62	15	58	33
Timothy	4	67	67	13	56	35

[1]Bermudagrass was grown in Georgia and the other grasses grown in Kentucky. All were harvested as regrowth in summer after an initial spring cutting.

Source: W.R. Windham, USDA Russell Research Center, Athens, GA.

desired chemical constituents (crude protein, neutral detergent fiber, acid detergent fiber) are not measured directly by instrumentation but rather from prediction equations developed from calibration samples analyzed by traditional chemical methods. This approach is increasingly being used by laboratories which serve livestock producers.

A comparison of the different forage quality measurements of several important perennial grasses is shown in **Table 16.4.** When comparing grass species it is important that they be harvested at the same stage of maturity or age.

5. Relative feed value. In recent years, ADF and NDF values have been useful in developing a relatively simple index to evaluate and compare forages. This index, called the relative feed value (RFV), is a prediction of the intake and energy value of a particular forage.

Table 16.5. **Quality standards for legume, grass, or grass-legume hay. Hay Market Task Force, American Forage and Grassland Council.**

Quality standard	CP	ADF	NDF	DDM	DMI	RFV
	---------------------			%	---------------------	
Prime	>19	<31	<40	>65	>3.0	>151
1	17-19	31-35	40-46	62-65	3.0-2.6	151-125
2	14-16	36-40	47-53	58-61	2.5-2.3	124-103
3	11-13	41-42	54-60	56-57	2.2-2.0	102-87
4	8-10	43-45	61-65	53-55	1.9-1.8	86-75
5	<8	>45	>65	<53	<1.8	<75

CP = crude protein, ADF = acid detergent fiber, NDF = neutral detergent fiber.
Digestible dry matter (DDM %) = 88.9-0.779 ADF (% of dry matter).
Dry matter intake (DMI) = 120/forage NDF (% of DM).
Relative feed value (RFV) calculated from (DDM x DMI)/1.29.
Reference hay of 100 RFV contains 41% ADF and 53% NDF.

Relative feed value compares or ranks forages according to their digestible dry matter (DDM) and potential dry matter intake (DMI). It actually is the relationship of DDM x DMI divided by a constant. Relative feed value is expressed as percent compared to full bloom alfalfa which has an RFV value of 100. The RFV values increase as forage quality increases (**Table 16.5**).

Digestible dry matter values used to calculate RFV are a percentage estimate of a feed or forage which is digestible. This is usually estimated from ADF values, but may be based on feeding trials with animals. Different equations are used with ADF values obtained for grasses, legumes, and sometimes for different regions of the country.

Dry matter intake (DMI) values used to calculate RFV are an estimate of the relative amount of forage an animal will eat. The DMI may be based on animal feeding trials, but usually is a calculation based on NDF values. As with DDM, different equations are used with different categories of forage.

In states where hay auctions are held, RFV has influenced the hay price more than any other factor in recent years. More and more producers are becoming oriented toward this term which allows a comparison of the feed value of various lots and/or types of hay.

Relative feed value is a good predictor of the ability of a forage to supply nutrients to livestock. The ADF and NDF values are important in determining the RFV of a forage. Acid detergent fiber values are used to calculate DDM; the lower the ADF, the higher the DDM. Dry matter intake is calculated using NDF values, and a low NDF results in high DMI.

While this system using RFV may seem confusing, it is just another way to numerically express a comparison of different forages. The same principles we have known for many years still apply: forage quality is best when fiber levels are low and digestibility is high. Furthermore, high intake by animals is indicative of high forage quality. High protein levels are desirable, but protein is not used in calculating RFV.

Relative feed value is a one-number estimate of the *energy* value of a forage, haylage, or silage. The RFV of mature (three-fourths to full bloom) alfalfa is equal to 100. A forage with an RFV of 125 has 25 percent more energy than mature alfalfa. Likewise, an RFV of 75 means a forage has 75 percent of the energy of mature alfalfa.

Relative feed value is not intended to be used to balance diets. Calculated energy (obtained from fiber values) and protein values will continue to be used for that purpose. However, RFV **is** of value in pricing hay because it provides a relative estimate of the feed value of hays (with regard to energy).

Summary

It is not necessary to understand how forage quality is measured in a laboratory, but some understanding of how forage quality affects animals **is** a fundamental requirement for obtaining good animal performance (see Chapter 17). A livestock producer needs to be aware of the major influences on forage quality such as stage of maturity, plant species and variety influences. It is the total quantity of available **nutrients** in a given amount of forage, and not the total quantity of forage, that is of primary importance in obtaining good animal performance. ■

Nutrient Requirements of Livestock

RUMINANTS have the unique ability to consume large quantities of forage and convert the fiber components (cellulose and hemicellulose) into useable forms of energy. Energy constitutes the largest nutrient requirement for ruminant livestock and is usually supplied by pastures, hay, silage, and sometimes grain. The most limiting factor in forage quality is digestible energy. Digestible energy is generally lower in perennial warm season grasses than in cool season grasses, and declines as plants become more mature (see Chapter 16).

All animals need a certain amount of protein in their daily diet. Energy requirements, which are more likely to be the primary concern on most livestock farms in the South, are higher for lactating and growing individuals than for mature animals. Most essential minerals are adequately supplied in forages. Calcium (Ca) and phosphorus (P) supplements may be needed when animals are consuming low-quality hay or grazing frosted warm season perennial grasses in winter (**Appendix A.27**).

Nutrient requirements of ruminant animals differ greatly, depending on class of livestock, weight, and desired daily production. Total digestible nutrient and crude protein requirements of several livestock classes are illustrated in **Table 17.1**. Forage species differ considerably in their digestibility and suitability as energy sources for different classes of cattle (**Figure 17.1**). For each group of forage species there is a range in digestibility, depending on stage of maturity and season of the year. Good pasture management is necessary to keep nutritive quality at the upper end of this range for best animal performance.

Table 17.1. Total digestible nutrient (TDN) and crude protein (CP) requirements of selected animal classes.

Animal class	TDN, %	CP, %
Growing beef steer		
450 lb (1.5 lb/day gain)	65	11-13
650 lb (1.7 lb/day gain)	68	10-11
Beef cow		
Lactating	60	10-12
Dry, pregnant	50	7-8
Sheep		
Lamb (finishing)	70	12
Ewe (lactating)	65	13
Ewe (maintenance)	55	9
Fallow deer		
Doe (lactating)	66	14-16
Growing buck	60-64	12-14
Meat-type goat		
Doe (lactating)	62	12
Growing buck	62-66	12-13
Horse (maintenance)	70	10-11

Source: M.A. McCann, Animal Science Dept., Univ. of Georgia.

Growing Animals

Growing animals such as steers or lambs have high requirements that can be met on small grains, annual ryegrass, or cool season

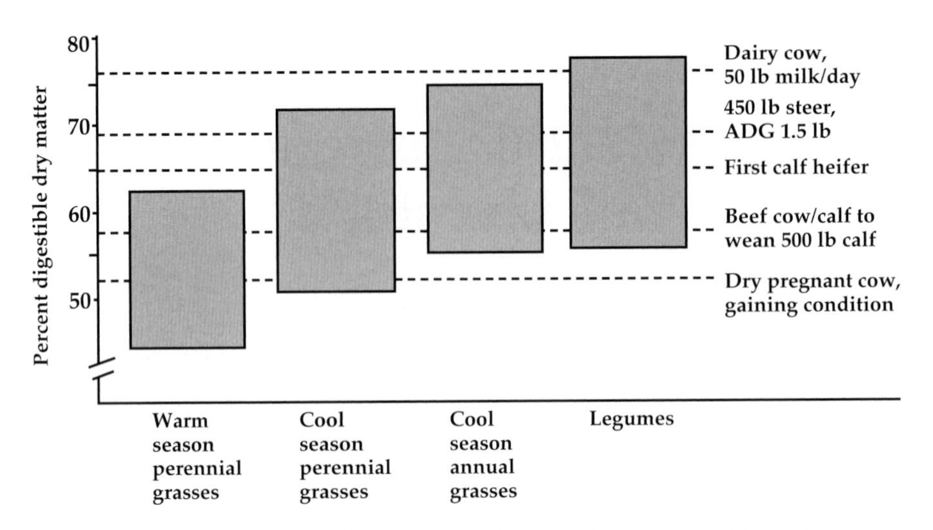

Figure 17.1. Forage digestibility ranges and their suitability for different classes of livestock.
Adapted from: H. Lippke and M.E. Riewe. 1976. Texas Agric. Exp. Stn. Res. Monograph RMGC: 169-206.

perennial grasses such as orchardgrass, Kentucky bluegrass, or endophyte-free tall fescue (see Chapter 27). Clovers, alfalfa, or birdsfoot trefoil companion crops can sometimes enhance animal gain on these grasses. Warm season perennial grasses such as bahiagrass or bermudagrass lack the quality for high gains and are usually unsatisfactory in profitable grazing of beef steers, lambs, and deer. Warm season perennial grass pastures can provide satisfactory gains for short periods of time in late spring and early summer if closely grazed to maintain a high leaf percentage, but animal performance declines sharply as the season progresses.

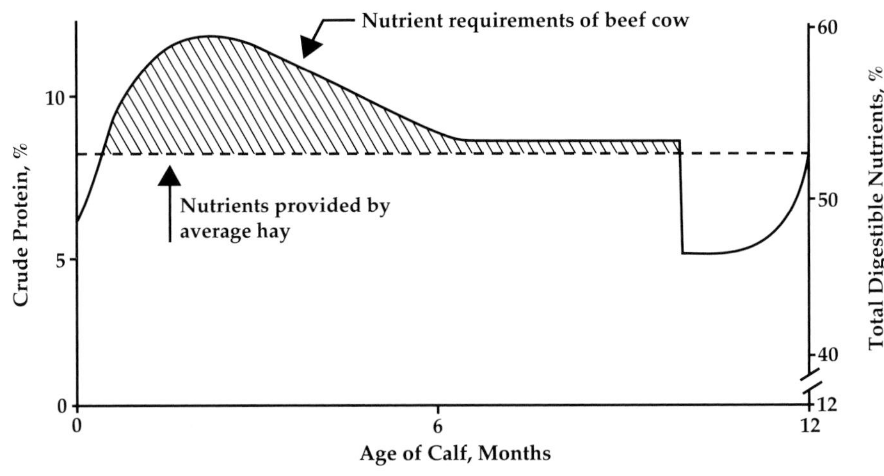

Figure 17.2. Average quality grass hay cannot supply the nutrient requirements of a beef cow for rebreeding and providing milk for the calf.
Source: M.A. McCann, Animal Science Dept., Univ. of Georgia.

Beef Cows and Calves

Most **growing** pastures of warm or cool season perennial grasses in the South will adequately supply the needs of a dry beef cow (see Chapter 26). However, after killing frosts, the quality of bermudagrass and bahiagrass declines rapidly, especially with heavy rainfall. Protein needs of dry pregnant beef cows are low, but protein supplements must be fed on frosted warm season grass pastures for adequate nutrition. Protein supplementation may also cause cattle to consume more of this type grass and utilize it more efficiently.

After calving, nutrient requirements of beef cows increase markedly for rebreeding and providing milk for calves. Until a calf is about 3 months old, it is mostly dependent on its mother for food. Calving during January and February occurs when perennial grass pasture growth is very limited, so good quality hay is important (**Figure 17.2**). Mineral supplements are necessary at this time. Lactating ewes have higher energy and protein requirements than beef cows, so it becomes even more important to have high quality pasture and/or stored feed for good lamb growth.

Winter-born calves begin to consume significant amounts of forage in spring when pasture quantity and quality are high. Pasture quality normally is lower in summer when cow milk production declines and the intake demands of calves are increasing. Overseeding pastures with legumes can improve forage quality and increase calf weaning weights. Creep feeding of grain to calves on pasture may increase weaning weights but is only profitable in times of low grain prices and/or high calf prices.

At high stocking rates, creep or forward grazing of beef calves can efficiently increase weaning weights of calves. For autumn-born calves in the lower South, planting cool season annuals on prepared seedbeds and stocking at up to 10 calves per acre have been efficient. For spring-born calves, simply allowing calves creep access to the next pasture ahead of their mothers in a rotational stocking method can efficiently increase weaning weights (see Chapter 25).

Replacement Heifers

Replacement heifers require special nutritional consideration. If heifers are bred to calve first as 2-year olds, they should be fed good forage or grain during the first and second winters similar to growing steers. Nutrients must be adequate for continued body growth, development of the fetus and the mammary system for feeding the new calf, and rebreeding. Ideally, high-quality winter annual or cool season perennial grass pastures can meet the nutrient needs of this class of animal. If grass hay is fed, it will generally need to be supplemented with protein and minerals. It is important that replacement heifers grow adequately but not fatten. Allowing heifers to become fat prior to calving is expensive and may cause calving problems.

Controlled Breeding Season

A controlled breeding season is essential in matching animal nutrient requirements to available pasture forage and reducing supplemental feeding. Generally this is most easily accomplished with autumn calving in the lower South, winter calving in the middle South, and early spring calving in the upper South. A controlled breeding season generally results in lower calf death losses, heavier weaning weights, and more efficient utilization of labor and resources.

Finishing

Although high quality forage will give good animal performance, grain feeding is needed to achieve maximum liveweight gain. However, since forage is a lower cost source of nutrients, it is usually more profitable to utilize high quality pasture rather than grain to put on as much muscle and frame as possible. When steers begin to fatten rather than grow (for British breed steers, between 800 to 1,000 lb), their requirements for digestible energy increase and cannot be met by warm season grasses. At this point, grain feeding is needed to achieve adequate finish in a reasonable time.

Grain feeding on high quality ryegrass, rye, or wheat pasture is usually not profitable. Higher stocking rates are possible with grain feeding, but there is little or no increase in individual gains as animals generally substitute grain for high quality pasture. While steers and heifers can adequately fatten on cool season grass and clover pastures, they are usually only marginally fat and will be discriminated against in price at slaughter. Therefore, it is typical in the USA for cattle to be fed on concentrate diets for at least 60 days and usually 120 to 140 days prior to slaughter.

Dairy Cows

Lactating dairy cows in the USA require large quantities of high-quality feed, including some grain to achieve optimum production per cow. High-quality forages, particularly alfalfa and corn for silage, are important for high milk production. During certain periods of the year some forages produce enough high quality forage to meet most of the demand for high levels of milk production by dairy cows. Dairy producers must also utilize stored feed of known quality to provide a diet to maintain lactation at a high level. High quality alfalfa hay or haylage and corn silage with grain supplements have been commonly used for dairy production in the USA (see Chapter 28).

Deer Farming

The fixed breeding season of most deer species means that fawns are born during June and lactation continues until weaning in September. The nutrient requirements to sustain lactation and provide rapid growth of fawns during this period are high and about double that needed in winter. High-quality alfalfa or clover pasture can reduce supplemental grain feeding costs.

Summary

For efficient and economical production, forage quality must be matched to the nutrient requirements of the class of animal being fed. Grazing beef cows with calves on expensive cool season annual pasture is inefficient unless they are limit-grazed (for example, 2 hours per day for 3 to 7 days per week). Grazing beef steers on low-quality pasture is unprofitable as it will result in low daily gains even though available forage and stocking rates are high.

Forage quality and quantity of pasture vary greatly during the year. The length of the growing season on adapted forage species differs considerably from the lower to the upper South. The nutritive requirements of animals in each area can be met by: utilizing a controlled breeding season to match quality and quantity of forage, supplemental feeding, and utilizing legumes in pastures. ■

Chapter 18

Hay Production

WHEN PASTURE growth is scarce, it is necessary to provide some type of stored feed for grazing animals. Hay offers a number of advantages: it can largely be mechanized; it stores well when adequately protected; and it can meet the nutritional requirements of most classes of livestock.

Hay fits well with a grazing program because accumulated forage can be cut for hay during periods of excess growth, thus minimizing waste. Harvesting excess growth also results in the subsequent production of young, tender forage regrowth for later grazing.

Hay Crop Selection

The first step in hay production is deciding what forage species to plant. This decision is obviously limited to the forage species adapted to the location. Other

Table 18.1. Approximate hay yield, crude protein, and total digestible nutrient (TDN) content of various hay crops.

Type of hay crop	Annual or perennial	Usual hay yield (tons/A)[1]	Approximate usual nutrient level[2] Crude protein, %	TDN, %
Cool season				
Alfalfa (early bloom)	P	3-6	17-22	57-62
Arrowleaf clover	A	2-3	14-17	56-61
Birdsfoot trefoil	P	1-3	16-20	57-62
Oats	A	1-4	8-10	55-60
Orchardgrass	P	2-5	12-15	55-60
Red clover	P	2-4	14-16	57-62
Rye	A	1-4	8-10	50-55
Ryegrass	A	1-4	10-16	56-62
Soybean	A	2-3	15-18	54-58
Tall fescue	P	2-4	10-15	55-60
Warm season				
Annual lespedeza	A	1-2	14-17	52-58
Bahiagrass	P	3-5	9-11	50-56
Coastal bermudagrass (4 wks)	P	5-8	10-14	52-58
Common bermudagrass	P	2-6	9-11	50-56
Dallisgrass	P	2-4	9-12	50-56
Johnsongrass	P	2-5	10-14	50-60
Pearl millet	A	2-6	8-12	50-58
Sericea lespedeza	P	1-3	14-17	50-55
Sudangrass	A	2-6	9-12	55-60

[1]Assuming the crop is grown in an area to which it is adapted using recommended production and harvesting practices.
[2]Dry matter basis, assuming recommended production and harvesting practices and no excessive weather damage. Forage quality is affected by many factors (see Chapter 16).

considerations should include nutritional needs of the animals consuming the hay, the relative quality of hay which can be produced from various adapted species, probable hay yield, and cost of establishment and maintenance of each of the crops which could be grown. The information presented in **Table 18.1** provides a basis for comparison of some species which can be grown for hay.

It is generally more economical to produce hay from perennials than from annuals due to avoidance of annual establishment costs. This is especially true when producing hay for animals which have relatively low nutritional requirements. Growing a legume/grass mixture for hay production will generally improve hay quality as compared to using grass alone.

When considering the economics of hay production, it is important to realize that a large proportion (usually 30 to 40 percent) of the expenses per acre involved in hay production are *fixed* costs. These are costs which will be essentially the same regardless of other factors. This provides incentive for striving for both high yield and high quality. If a livestock producer is going to expend time, energy, and money in producing hay, it is worthwhile to make it an efficient operation in which high production and excellent quality hay are obtained.

Producing the Forage

Perennial hayfields are normally in use for many years, and a poor start may have a negative impact for years to come. The goal should be a thick, vigorous stand of an adapted, productive forage variety. To achieve this, it is necessary to plant an adequate quantity of high quality, weed-free seed or vegetative material of a recommended variety. Recommended establishment practices as discussed in Chapters 11 and 12 should be followed.

Since yield is important in obtaining economical hay production, it is essential to lime and fertilize hayfields according to soil test recommendations. Weeds and insects should be controlled with pesticides as appropriate.

Some livestock producers do not fully understand what they should be striving for when producing hay. On many farms, the emphasis is generally on the quantity of hay produced per acre. The most important consideration, however, is not the quantity of **hay** produced, but rather the quantity of available and digestible **nutrients** stored and the level of animal performance which will result from feeding those nutrients.

The two most important criteria used in evaluating hay quality are the crude protein content and the energy value (see Chapters 16 and 17). Although both are important factors, low digestible energy is usually the main limiting factor in Southern livestock rations. Therefore, the emphasis with regard to forage quality of hay should generally be on improving the digestible or available energy value.

Factors which can influence hay quality include: plant species, plant variety, weeds, insect damage, diseases, weather at harvest, and harvesting techniques. However, hay quality is most likely to be affected by two other factors, both of which are under the control of the producer: (1) fertilization; and (2) stage of maturity at harvest.

Nitrogen (N) fertilization will, up to a point, increase the protein content of grass hay. Fertilization with other nutrients such as potassium

(K), phosphorus (P), magnesium (Mg), and sulfur (S), may also influence the amounts of these elements which are present in forage.

Periodic soil testing followed by applying the recommended nutrients will allow the levels of specific nutrients in forage to be adequate for animals. Fertilization may also improve hay palatability to animals and thus influence animal performance by increasing intake. Although important in obtaining good hay yields, fertilization normally has little or no influence on the energy level of hay except by encouraging the domination of desirable forage species.

The single most important producer-controlled factor influencing hay quality is stage of maturity at harvest. This is where many Southern livestock producers can most easily and dramatically improve hay quality.

The stage of maturity at harvest influences the palatability, crude protein content, and especially the digestible energy level. In general, the best time to harvest for a good yield as well as high energy and crude protein levels is in the early bloom stage for legumes and in the boot stage (just before seedhead emergence) in grasses. Exceptions to this rule are: sericea lespedeza–15 to 18 inches; summer annual grasses–30 to 40 inches; and hybrid bermudagrass–4 to 5 week intervals or 15 to 18 inches. Forage quality deteriorates rapidly with advancing maturity even though yield will continue to increase. Advice regarding specific times to cut various forages is provided in **Table 18.2.**

Table 18.2. Recommended stages to harvest various hay crops.[1]

Plant species	Time of harvest
Alfalfa	Bud stage for first cutting, one-tenth bloom for second and later cuttings. For spring seedings, allow the first cutting to reach mid- to full bloom.
Orchardgrass, timothy, or tall fescue	Boot to early head stage for first cut, aftermath cuts at 4 to 6 week intervals.
Red, arrowleaf, or crimson clovers	Early bloom.
Sericea lespedeza	Height of 15 to 18 inches.
Oats, barley, or wheat	Boot to early head stage.
Soybean	Mid-to-full bloom and before bottom leaves begin to fall.
Annual lespedeza	Early bloom and before bottom leaves begin to fall.
Ladino clover or white clover	Cut at correct stage for companion grass.
Hybrid bermudagrass	15 to 18-inch height for first cutting, mow every 4 to 5 weeks or when 15 inches high.
Birdsfoot trefoil	Cut at correct stage for companion grass.
Sudangrass, sorghum-sudan hybrids, pearl millet	Height of 30 to 40 inches.

[1]Adapted from: J.D. Burns, J.K. Evans and G.D. Lacefield, "Quality Hay Production", Southern Regional Beef Cow Calf Handbook, SR 5004. (Appendix A.26).

If a producer waits until past the recommended stage to cut hay, the fiber content increases and palatability and digestibility decline. Waiting until later will increase the number of bales or tons of hay produced, but nutritive value decreases.

In addition to requiring more fuel, time, and labor to store the hay, the farther past the optimum stage that hay is harvested, the poorer animal performance will be because of low digestible energy and high fiber. Poor quality hay passes more slowly through the animal digestive system, causing lower intake of low quality hay, which further reduces animal performance.

Thousands of tons of hay have been rain damaged or harvested too late because of equipment problems. Before beginning the harvesting process, all hay equipment should be checked and serviced. This should include sharpening knives, checking belts, and lubrication.

Some hay crops require that a particular stubble height be left. For example, sudangrass, sorghum-sudan hybrids, and pearl millet should not be cut lower than 6 to 8 inches. Sericea lespedeza should not be cut lower than about 4 inches.

Hay should normally be cut in late morning, afternoon, or early evening when the dew has dried off the forage. It is important that the mower be set to cut forage at the proper height. The travel speed of the tractor should be slow enough that all the forage is cut cleanly and evenly. Periodic adjustments in speed may be necessary due to variability in stand thickness.

Many producers use mower/conditioners or have hay conditioners which are used separately from mowers. Conditioners (or crimpers) can reduce drying time by as much as 50 percent. They are essential for thick-stemmed forages such as pearl millet and sorghum-sudan hybrids.

"Respiration" refers to the breakdown of food materials within plants. This process, which occurs in all living plants, continues after forage plants are cut and until the moisture content drops to below about 40 percent. Respiration can result in dry matter losses of 2 to 16 percent, depending on the situation. When drying conditions are good, respiration losses are minimized; when drying conditions are poor, respiration losses may be high.

Rain can damage hay in several ways. It leaches soluble nutrients, keeps the moisture level high, prolongs the period of respiration loss, and increases the likelihood of mold.

By crushing plant stems, a conditioner causes moisture levels to drop much more quickly to the 40 percent level at which respiration losses cease. An additional benefit is that when hay dries more quickly, the likelihood of rain damage is reduced.

Leaf losses are common with legumes such as alfalfa, sericea lespedeza, and red clover. When legume hays are raked or tedded at low moisture levels, leaf losses can be high. Leaf losses of 5, 10, and 20 percent may occur at moisture levels of 50, 35, and 20 percent, respectively. Therefore, it is advisable to complete raking before legume hays fall below 40 percent moisture.

The usual moisture range for raking should be around 45 to 55 percent. To reduce leaf loss, hay should be raked in the same direction as it was mowed. It may be desirable to rake legume hay early in the morning before the dew has dried in order to minimize leaf shatter.

Raking should be done with great care because this is the greatest source of leaf loss during harvest. Windrows should not be too thick because this reduces the amount of hay exposed to the air. Narrow, thick windrows will not dry as rapidly as wide, thin ones. Also, hay dries more quickly in a swath than in a windrow.

Hay drying may also be speeded with commercial hay drying agents (dessicants), the active ingredient of which is usually potassium carbonate. Such products are normally applied with a spray boom mounted just ahead of the mower with a deflection bar in front of the boom. The bar bends the plants over and allows good spray coverage. Dessicants break down the waxy coating on the stems of certain legume hay crops, principally alfalfa, and can be of benefit on some farms.

Baling Hay

Baling should also be done in the same direction as mowing and raking. Baling should progress at a slow enough speed that the hay will be cleanly and evenly fed into the baler. It is important to keep the density, size, and shape of bales relatively constant as this aids in storage and handling.

Leaf losses can also be high during baling operations, ranging from 1 to 15 percent. One way to reduce leaf loss at baling is to spray an organic acid (propionic acid is most commonly used) on the hay which then allows baling at moisture levels up to 30 percent rather than the 15 to 20 percent range normally required for safe storage without mold damage.

Baling losses with conventional balers (producing small rectangular bales) are typically 3 to 8 percent, while baling losses with large round balers can vary from 5 to 15 percent. Thus, the **potential** baling loss is greater with large round balers, but a skillful operator may keep such losses as low or lower than would be obtained with a conventional baler.

It should be emphasized that these figures do not account for storage or feeding losses (see Chapter 19). Round bales stored outside may have a high spoilage loss, while hay stored inside should have virtually no storage loss. Feeding losses also tend to be much higher with round bales than with conventional bales.

Forage Testing

The best way to determine the nutrient content and value of hay is to have a sample analyzed at a testing facility (see Chapter 16). To obtain accurate results from such tests, a good representative sample must be obtained. The best means of getting a good sample for testing is with a core sampler. Once the results of a test are obtained and the quality of a lot of hay is known, that information can be used in effectively formulating diets for livestock. ■

Figure 18.1. A skillful operator can keep baling losses low with large package hay equipment.

Chapter 19

Hay Storage and Feeding

POOR TECHNIQUES of hay production can lower hay yield and quality. However, substantial losses can also occur in both storage and feeding processes. Livestock producers need to be aware of, and strive to avoid, all potential forage losses.

Storage

Small rectangular bales of hay should be protected from the elements. Otherwise, weathering losses will be high because of the relatively high level of exposed surface area. Normally it is best to store rectangular bales inside a building.

Stacks of rectangular bales can be stored outside and covered with plastic or a tarpaulin, but this is usually unsatisfactory. With time, wind often partially or totally uncovers the hay. Furthermore, moisture often condenses under the cover at the top of the hay stack, resulting in spoilage. This is particularly true if plastic is used to cover hay.

Large round bales and stack systems are popular storage options. One of the selling points for large package balers is that the bales can be stored outside. However, there are times when this may be unwise.

Large-stemmed grasses such as pearl millet and sorghum-sudangrass hybrids, and crop residues do not form a tight bale, and are easily penetrated by water. In addition, high quality hays such as alfalfa are so valuable that even a relatively small percentage of spoilage is costly and should not be tolerated.

Furthermore, legume hays stored outside sustain more damage than most grass hays. With such hay it is usually best to minimize losses by storing it inside. If this cannot be done, such "high risk" or high quality hay should be fed early in the season to minimize losses.

Figure 19.1. Great differences can occur in the amount of hay wasted during feeding, depending on management.

Inside storage is best, regardless of the type of hay package, so round bales should be stored inside if space is available. Once hay is protected from the elements, it loses little nutritional value even if it is stored for several years. However, it **will** change color and become visually less attractive. Sometimes it is not feasible to store large packages of hay inside. In such cases, round balers should normally be used only for putting up types of hay which form a tight bale that readily sheds water (grass or grass/legume).

When large round bales or hay stacks are to be stored outside, selection of a storage site is an important consideration. A hay stackyard should be located in an open, well-drained, sunny location convenient to feeding areas. It is particularly important to select a well-drained spot.

Much of the spoilage in large hay packages stored outside typically results from moisture being absorbed from the ground rather than penetrating from the top. Thus, it is desirable to reduce or eliminate hay/soil contact if possible. This can be done by storing bales on old tires, railroad ties, crushed rock, or a concrete pad.

Large bales or stacks should be promptly moved to a storage site to prevent killing spots of grass in the hayfield by shading. It is best to leave a space of about 18 inches between bales to allow good drying after rains. Storing hay bales end-to-end (flat sides together) is acceptable, but storing them with the rounded sides touching is not advisable due to the creation of a site where moisture can accumulate.

If placed on a hill, rows of bales should run up and down the slope so they do not act as a dam for surface water. Stacking round bales on top of each other is an undesirable practice unless the stack is to be covered. There may be some benefit to storing bales with the flat sides facing north and south. Theoretically, this allows sunshine to help keep the bales dry a higher percentage of the time.

Large round bales should not be stored near wire fences or other objects which may attract lightning. To further minimize fire loss risk, it is desirable to store hay in more than one location. Grouping hays of similar quality together in individual stackyards can facilitate efficient feeding. Also, a road next to a stackyard may serve as a firebreak.

It is advisable to have more hay on hand than one expects to feed. Winters that are longer than expected, or summer droughts, can often cause severe problems for producers who are not prepared. The minimum length of the winter feeding period varies depending on location, pasture species available, and weather. For a livestock producer in Zone A who plants winter annuals in early fall, the winter feeding period could be less than 30 days; on the other hand, it may be more than 120 days in Zone D.

A dry beef cow will consume at least 15 lb of hay per day while a lactating cow may consume 25 to 30 lb or more per day. A "rule of thumb" is that ruminants will usually consume 2.5 to 3 percent of their body weight in dry matter per day. Horses will consume 1 to 2 percent of their body weight in dry matter per day. By using these figures, one can calculate the approximate quantity of hay which might be needed to overwinter a group of animals and still have enough left over for emergency needs (see Chapter 26 and **Appendix A.19**).

Heating and Fire Hazard

Hay baled at too high a moisture level will heat. In addition to causing reduced digestibility of protein and hemicellulose, the hay may catch on fire. Because of the possibility of fire, fresh, green hay should never be stored tightly against older, dry hay.

Hay temperature can be monitored by making a probe out of a 10-foot piece of 2-inch diameter pipe on which one end has been sealed with a sharpened plug. The pipe can then be driven into a stack or large bale of hay followed by lowering of a thermometer into the pipe. Readings should be taken at several locations and depths, leaving the thermometer at each location for 10 to 15 minutes. When hay temperatures remain below 120° F, it is considered safe; between 120 to 140° F is considered a danger zone and the hay should be closely watched; if the temperature goes above 160° F, a fire is likely.

Hay which is heating to an unacceptable, dangerous level should be moved to a spot where fire will not destroy anything except the hay. The danger of fire will generally subside within 2 to 3 weeks.

Feeding

Feeding losses occur mainly from trampling, refusal, and leaf shatter. Some feeding loss is inevitable, but it can vary from as little as 2 or 3 percent to well over 50 percent. Since hay production is a major expense in most livestock operations, a goal should be to develop a feeding system which minimizes these losses.

The traditional method of feeding, although labor intensive, was to feed rectangular bales in a manger or bunk, or to transport them to a sod area for feeding. When only a one-day supply of hay is offered at a time, feeding losses are minimal.

Some balers produce small, round bales which may be left in the field and fed in place. This system works fairly well, but only if feeding losses are minimized by allowing the animals access to only a few days' supply at a time. Otherwise, feeding losses will likely be high.

The first rule of feeding hay is to prevent unlimited access, thus reducing losses. This can be accomplished by moving one or a few bales at a time from the stackyard to an area where they will be fed. Use of feeding racks or panels can also greatly reduce losses (**Table 19.1**).

The losses from feeding hay as bales and as loose stacks were com-

Table 19.1. Effect of feeding systems on losses of Johnsongrass hay and on steer performance.

Item	Conventional bales on sod	Round bales on sod	Round bales with panels
Hay fed, lb/day	9.1	19.1	12.3
Days on test	79	79	79
Initial wt. (lb)	535	538	538
Final wt. (lb)	615	635	646
Gain/animal	80	97	108
Average daily gain (lb)	1.0	1.2	1.4
Lb hay required/lb of gain	9.0	15.6	9.0

Adapted from: W.B. Anthony, E.S. Renoll, and J.L. Stallings. 1975. Alabama Agric. Exp. Stn. Cir. 216.

pared in east Texas[1]. In this work, 20 cow-calf units and a bull were group-fed baled hay or stacked hay. Cattle in the stack hay group were allowed access to one stack at a time. Baled hay was fed on the ground, and the number of bales fed was adjusted so that consumption was completed in a 2-day period. Waste from the baled hay groups was approximately 3 percent. Hay waste from the stack hay group ranged from 18 to 25 percent.

When stacks were enclosed by stack guard panels, hay losses were less than 5 percent. However, it is important to note that labor and fixed costs associated with protecting hay may partially offset the losses to waste.

Other methods of limiting animal access such as an electric wire or a gate may also be helpful. The point to remember is that **to minimize losses, access of the animals to the hay must be limited.** In situations in which there is no feeding rack, panel, or other restrictive device, it is important to feed only as much hay as the animals can eat in one or two days. A "rule of thumb" is to provide 1 foot of access per cow to large hay packages being fed without feeding racks or panels.

If hay is fed on sod, the feeding areas should be rotated. This minimizes soil compaction and sod kill and also causes dung and urine to be more evenly distributed. Therefore, it is desirable to select less fertile fields or areas when feeding on sod. Feeding the lowest quality hay first is a desirable practice. The refused hay, which may be a substantial quantity, helps provide a footing for further feeding.

There is less refusal and waste when high quality hay is fed, thus further emphasizing the importance of striving for high hay quality. Another approach used by some producers is to force animals having low nutritional requirements to

[1]F.M. Rouquette, Jr., and N.K. Person. 1978. Texas Research Monograph RM6C, Texas Agric. Exp. Stn.

Figure 19.2. Storing round bales of large-stemmed or high value hay under shelter can save money.

Figure 19.3. Feeding round hay bales from a bale wagon which can be moved to different pasture areas will result in better distribution of dung and urine.

clean up a feeding area before more hay is fed.

Other Possibilities

Net wrap, plastic wraps of various types, sleeves, and tarps can all be used to reduce hay spoilage losses. These treatments may or may not be cost effective, however, depending on the vulnerability and value of hay to be stored, length of the storage period, and amount and seasonal distribution of rainfall in the area.

Barn drying is appealing from the standpoints of preserving hay quality and minimizing the risk of rain damage. The problems with barn drying are building an appropriate facility and (especially) economically providing the energy for drying. While barn drying is an intriguing concept, it does not appear likely to become popular unless a low-cost system of providing energy for drying is discovered.

The cost of building a shed for storing round bales varies considerably within the region, as does the length of time for which storage may be needed. In addition, the amount of space needed for storage is greatly affected by bale density. For example, it requires about half the space to store high density hay which may contain 9 pounds per cubic foot as for hay baled at a low density of 5 pounds per cubic foot.

Because the feasibility of using various techniques for protecting hay varies greatly, it is difficult to make general recommendations. Producers should rely on local experts and their own calculations when making decisions regarding hay storage techniques.

Summary

Many of the points important in storing and feeding hay are common sense. Nonetheless, much hay is wasted. Losses are often more than a producer might think. The losses resulting from improper storage and poor feeding practices are particularly objectionable because the expense of producing the hay has already been incurred. Performing these steps correctly is worth the effort. ■

Chapter 20

Silage Production and Feeding

SILAGE is defined as plant material that has undergone fermentation in a silo. A silo is any storage structure in which green, moist forage is preserved.

Harvesting, preserving and feeding forage crops as silage is a practice used throughout the South. Silage represents a convenient and economical source of feed for the dairy industry and is increasingly used in beef backgrounding and finishing programs.

Silage has many advantages, including: 1) less field and harvest losses than occur with hay (**Figure 20.1**); 2) choice of many crops which can be used for silage; 3) mechanization of harvesting the standing crop to feeding; 4) less likelihood of weather damage during the harvesting process; 5) when properly ensiled, it can be stored for long periods with only small losses of nutrients; and 6) it can be flexibly used in many livestock feeding programs. Obviously, these advantages should not be ignored. It is likely that many Southern livestock producers could benefit by incorporating silage capability into their operations.

Some disadvantages of using silage include: 1) bulky to handle and store; 2) requires additional equipment and structures for harvesting, storing and feeding; 3) excessive losses if not stored properly; 4) not readily marketable if not used on the farm; and 5) must be fed soon after removal from the silo to minimize spoilage.

Crops for Silage

Many crops grown in the South can be successfully preserved as

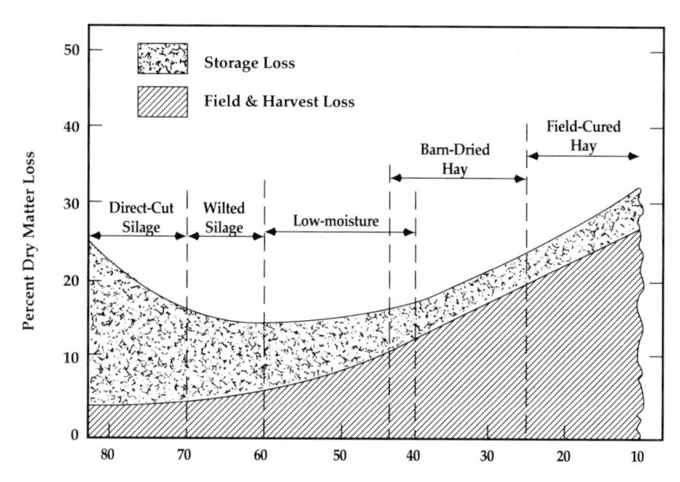

Figure 20.1. Estimated total field and harvest loss and storage loss when legume-grass forages are harvested at varying moisture levels and by alternative harvesting methods.
Source: C.R. Hoglund. 1964. Michigan Agric. Exp. Stn. Agric. Econ. Publ. 947.

silage. The type of livestock to be fed, expected yields, machinery available, soil types, and planting sites can enter into the decision as to what crops to plant for silage.

Since most of the above-ground parts of the crop are removed from the field during silage making, removal of fertilizer nutrients is much greater than when the same crop is harvested for grain. Necessary adjustments should be made in the fertility program to insure adequate fertilizer nutrients for the following crop.

High-energy crops (corn, grain sorghum, small grains) have been used extensively for silage (**Table 20.1**). Other forage crops (alfalfa, forage sorghum, legume-grass mixtures and various grasses) are also used, but require wilting to reduce moisture content for proper ensiling. Corn ranks as the best silage crop. Corn silage is high in energy and results in good animal performance.

Grain sorghum silage made from grain-type sorghum is slightly lower in quality than corn silage. Silage yields from true grain sorghum varieties are also normally lower than silage yields from corn.

Forage sorghum is widely grown for silage. As with grain-type sorghums, forage sorghum results in somewhat lower animal performance than does corn silage, but yield per acre can be high. Forage sorghum varieties have widely varying characteristics. For example, time from planting to harvest varies from 80 to 130 days or more.

The ratio of sorghum grain to forage varies from 6 percent (forage types) to 35 percent or more (intermediate types), and yields vary from 10 to 30 tons per acre (wet basis) from a single harvest. In the extreme lower South (Zone A), it is possible to obtain two silage harvests from a single early planting of grain or forage sorghum. Considerable variation in yield and quality exists among sorghum varieties.

Summer annual grasses (pearl millet, sorghum-sudangrass hybrids), small grains, alfalfa, Johnsongrass, and surplus pasture growth of many other forages can also be harvested for silage. When cut at the right stage and properly ensiled, grass silage quality may vary from poor to good.

High moisture grass silage, cut and directly stored, makes poor quality silage unless an energy source is added. Warm season grasses such as Coastal bermudagrass are difficult to preserve as they have low available energy. Ground corn added at the rate of 100 to 150 lb/ton of green weight will facilitate bacterial fermentation and thus improve the quality of such silage.

The Ensiling Process

Many changes occur once green moist forage is placed in a silo. The speed with which these changes occur and the length of time required for ensiling vary with type of crop, moisture content, length of chop, type of silo, etc. If conditions are favorable the entire process will be complete within 20 to 21 days, resulting in a stable, high quality, pleasant-smelling silage.

Although many changes occur during the ensiling process, the overall goal is to use the oxygen and lower the pH so the forage material can be pickled or preserved. To accomplish this, two groups of bacteria are essential. The first group are oxygen-using (**aerobic**) bacteria which give off carbon dioxide and heat. If the silage is packed well and chopped fine, these bacteria will utilize all the oxygen in 4 to 6 hours.

The heat given off in the process will raise the temperature to 80 to 100° F.

The second group of bacteria (**anaerobic**) take over and work in the absence of oxygen to produce acetic acid. After the second or third day, lactic acid bacteria become active. Production of lactic acid occurs for 16 to 18 days, until the pH drops to between 3.6 and 4.2. At this low pH (high acid) the silage is pickled and all bacterial activity stops. At this point, the silage is stable and will keep for long periods of time, providing oxygen does not enter the silo.

Knowledge of the basic ensiling process helps us understand and appreciate the importance of management practices. Fine chopping and good packing are important to minimize the amount of oxygen which must be eliminated by bacteria before acid production. Having a readily available supply of digestible energy for the acid-producing bacteria greatly favors a good ensiling process. This is why corn and other high energy grain crops are excellent silage crops.

The Silo

Any storage structure in which green moist forage is preserved can be referred to as a silo. In addition to preserving the silage, the silo must protect the silage from damage by air, water, birds, rodents, and other animals during storage. Silos are divided into two basic types–upright and horizontal. Upright silos include conventional (concrete stave) and oxygen-limiting. Oxygen-limiting have higher original and operating costs than conventional, but offer advantages including bottom unloading, lower storage losses, more flexible use allowing constant refilling, and storage of more than one crop per season.

Horizontal silos include trench (below ground), bunker (above ground with sides), and stacks (above ground without sides). Trenches and bunkers are intermediate in cost to upright and stack silos. Horizontal silos have higher storage losses than upright. Packing is critical with horizontal silos. During

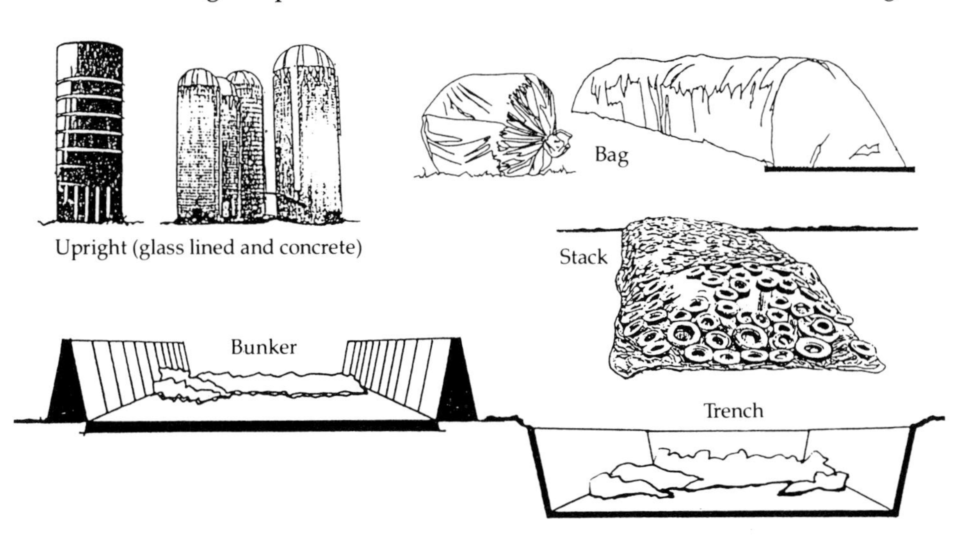

Figure 20.2. Illustrations show different types of silos.

filling, silage must be packed constantly. Top spoilage can be excessive if the silage is not covered and sealed to keep out air and water. Packing and sealing are more difficult with stacks since they have no permanent sides. Stacks are usually used for temporary storage or in emergency situations. More losses are usually associated with silage in stacks as compared to other silo types.

Equipment is now available that permits silage to be harvested, packed into plastic bags, and ensiled. Length of the plastic (silo) can vary to meet individual needs. In addition, plastic bags can be used for non-chopped forage. Forages can be cut, wilted and harvested in large round bales at high moisture content, placed in individual plastic bags, tied, and sealed.

Hay-crop silage (haylage) can be harvested and stored in upright airtight silos at a lower moisture content (45 to 55 percent). Less fermentation occurs with low-moisture hay-type plants which contain fewer readily available fermentable carbohydrates so silage is not fully "pickled" or preserved.

Alfalfa or alfalfa-grass can be an excellent silage crop. Interest in and use of alfalfa as silage is increasing in the South. Alfalfa can be harvested for silage over a wide range of moisture depending on individual operation and type of silo used for storage.

Factors Affecting Silage Quality

Many factors affect the quality of silage available at the time of feeding. Type of crop stored, stage of maturity, moisture content and length of chop are all important. With any crop grown for silage, the stage of maturity when harvested has the greatest effect on quality. Guidelines for harvesting various forage crops are provided in **Table 20.2**. As silage plants age or mature, quality declines, resulting in lower animal performance.

Table 20.1. Annual yield and composition of silage crops

		Yield, tons/A[1]		Dry matter basis	
Crop	Stage	Dry matter	35% dry matter	Crude protein, %	Crude fiber, %
Corn	Hard dough	6	16	8	24
Grain sorghum	Dough	4	11	9	28
Forage sorghum	Dough	5	14	8	33
Forage sorghum	Early head	4	12	11	29
Sorghum-sudan	Dough	4.5	13	10	34
Sorghum-sudan	Early head	2.5	7	12	26
Barley	Dough	2.5	7	9	27
Wheat	Dough	2.5	7	9	28
Oats	Dough	2.0	5	10	32
Rye	Boot	2.0	5	13	33
Alfalfa	Early bloom	4	11	18	31
Red clover-orchardgrass	Early bloom	2.5	7	12	33

Source: S.J. Donohue, R.L. Harrison and H.E. White (Ed.). Handbook of Agronomy. 1984. Virginia Agric. Ext. Ser. Pub. 424-100.

[1]Yields vary with location, climatic conditions, fertilization, and other factors.

Table 20.2. Usual recommended stage of maturity for silage harvest.[1]

Crop	Stage of maturity
Corn	Kernels will be dented and black layer visible.
Grain sorghum	Late milk to late dough.
Forage sorghum	40 inches or late boot stage.
Sorghum, sudangrass, Johnsongrass, millet	40 inches or boot stage whichever comes first.
Small grains, ryegrass	Boot to early head.
Soybeans	Late bloom–seed forming in pods and before lower leaves fall.
Alfalfa	Bud to early bloom.
Cool-season grasses	Boot to early head-first cutting; afterward 4 to 6 week intervals.
Hybrid bermudagrass	15 inches at first harvest, thereafter at 4 to 5 week intervals.

[1] These are guidelines. The quality of feed needed for the type or class of livestock to be fed should also be taken into consideration.

Moisture Content

Depending on the crop ensiled and the type of silo used, silage can be made over a wide range of moisture contents. If the crop is harvested at a stage when the silage is too dry, extra water must be added to insure good packing and proper fermentation. Low moisture may allow bacteria, yeast, and molds to become active, thus increasing the silage temperature. Overheating lowers feeding value and increases dry matter loss. Overheated silage will usually be brown to black with a carmelized (burnt sugar) odor. Extreme cases of overheating can result in spontaneous combustion.

If silage is harvested too wet, excessive seepage will result in loss of nutrients. High moisture may also result in silage being packed so tightly that all the oxygen is used up before the temperature gets to 80° F. Lactic acid bacteria are usually not very active below 80° F. Undesirable (butyric acid) bacteria become active, use carbohydrates and give the silage a foul odor and a taste like rancid butter. Other organisms can break down protein, giving the silage an ammonia odor. This results in unpalatable silage of low quality with a low feeding value.

When using stacks, trenches or bunkers, the moisture content of the silage should be about 65 to 70 percent in order to obtain adequate packing. Moisture content can be determined by using a moisture meter or by drying and weighing samples. A field technique for **estimating** moisture content is offered in **Table 20.3.**

Table 20.3. Field technique for estimating moisture content of chopped forage.

Condition of forage ball	Approximate moisture content
When the ball holds its shape and there is considerable free juice	Over 75%
When the ball holds its shape but there is very little free juice	70 to 75%
When the ball falls apart slowly and there is no free juice	60 to 70%
When the ball falls apart rapidly	Below 60%

Figure 20.3. Silage should be chopped at a length of 3/8 to 1/2 inch.

Length of Chop

Excluding air is the first, and a critically important, phase of the ensiling process. To insure that this phase is short, it is important to exclude air by fine chopping and packing adequately. The silage should be chopped to a length of approximately one-half inch. This helps insure proper fermentation by releasing plant juices and permits better packing for oxygen exclusion.

Silage Preservatives and Additives

Several materials have been investigated as to their merits for improving silage. Some products have proven to be helpful in certain situations. For example, adding anhydrous ammonia to corn silage, either as it is being chopped in the field or as it is being blown into a silo, is an effective and economical way of increasing protein content. Products such as enzymes, yeast cultures and antibiotics continue to be investigated. Such products have not proven to be effective and/or economical. Acid-forming bacteria have shown promise, especially with early spring and late autumn cut haylage. Research into the complex microbial relationships involved in silage making continues.

Careful consideration should be given in selecting silage additives, however. No additive can substitute for basic silage-making principles such as harvesting at the proper stage of maturity, storing at the proper moisture content, chopping finely, packing well, and covering adequately. Silage can be no better as a feed than the forage put into the silo.

Silo Gases

Lethal gases may occur in upright silos at any time during and soon after filling, but the greatest danger is from 12 to 72 hours after filling. Silo gases may be present in any ensiled material grown on any type of soil under any level of fertilization.

Two of the most dangerous gases may be recognized by their irritating odor and color. Nitrogen dioxide is reddish brown and nitrogen tetraoxide is yellow. A third gas, nitric oxide, is colorless and may be present and undetected at lethal concentrations. A few simple precautions will prevent tragedy or injury from silo gases:

(1) The blower should be run for 15 to 20 minutes before anyone goes into a partly filled silo. It should be left running as long as anyone is inside.
(2) No one should enter the silo for at least one week (preferably two weeks) after it has been filled.
(3) The silo room should be ventilated for at least two weeks after filling.
(4) The doors between the silo room and barn should be kept closed to protect livestock.
(5) If a person in the vicinity of the silo experiences the slightest throat irritation or coughing, they should get into fresh air immediately. Immediate treatment by a physician is essential.

Figure 20.4. Corn silage is a mainstay in dairy production.

Feeding Silage

Silage is ready to be fed 3 to 4 weeks after it is stored. Silage can supply some or all of the needs of various classes of livestock. Feeding can be mechanized for convenience and efficiency. In order to have an efficient silage feeding program, sufficient bunk space must be provided to accommodate the animals.

Likewise, nutritional needs of the animals must be known. Once the animal class and desired level of performance are determined, the amount of silage along with needed supplementation can be determined. To determine the level of supplementation needed, quality of the silage must first be determined. A forage (silage) sample should be collected and analyzed for nutrient levels.

Considerable variation can exist in silages. In general, corn silage is low in protein and must be supplemented for growing or milk-producing animals. On the other hand, alfalfa and other forage crop silages may be low in energy. A combination of high-energy (corn or sorghum) and high protein (alfalfa) has the potential for nearly balanced, home-grown rations.

Feeding losses are usually low in silage feeding programs. Feeding daily the amount that will be consumed prevents unused silage from spoiling in the bunk. This is especially important during warm weather.

Summary

Silage is an important source of stored feed in the South. It can be made from many crops and used in many different animal feeding programs. Silage represents a convenient and economical source of feed for the dairy industry and is increasingly used in beef backgrounding and finishing programs. ■

Chapter 21

Common Forage-Livestock Disorders

TOXIC SUBSTANCES or mineral imbalances in forages and weeds consumed by livestock may result in reduced productivity, visible symptoms of ill health, or even death of grazing animals. In the South, one of the most common ailments is fescue toxicity (see Chapter 23). Several other disorders are also potential problems for Southern livestock producers.

Bloat

Bloat can be a serious problem in cattle grazing pastures dominated by certain legumes. Other ruminant animals can be affected, but appear to be less susceptible than cattle. Bloat is caused by formation of a stable foam in the rumen which prevents eructation (belching) of gases produced by microbial fermentation of forage. As a result, the gases normally lost by eructation are retained. The pressure in the left side of the rumen then increases, resulting in the eructation mechanism being inhibited. When this occurs, the animal's oxygen supply is reduced or cut off, eventually causing it to suffocate. Animals affected will swell rapidly, and in acute cases, death can sometimes occur within an hour. Individual animals differ in susceptibility to bloat (**Figure 21.1**).

Persian, ladino, or white clovers, and alfalfa are examples of plants which have bloat potential. Certain legumes, including birdsfoot trefoil, sericea lespedeza, annual lespedeza, crownvetch, arrowleaf, and berseem clovers rarely or never cause bloat. This is probably because leaf tannins within these plants act as protein precipitants which aid in breaking up the stable foam in the rumen. Likewise, tropical legumes such as kudzu, cowpea, perennial peanut, and alyceclover rarely or never cause bloat. Occasionally, bloat occurs on lush pastures of ryegrass or small grains, usually in spring. Feedlot bloat in beef cattle raised on high-grain diets that may or may not contain legume forage occurs rather infrequently.

Control. Hungry animals should not be turned into a lush legume or winter annual grass pasture. Dry

Figure 21.1. Animal in early stages of bloat. Note that the animal's left side is especially distended.

hay should be fed before allowing animals to graze such a pasture. Limited hay feeding during the initial days of grazing can also reduce the incidence of bloat. Cattle placed on legume pastures should be checked frequently and removed quickly if bloat develops. The first signs of bloat will be swelling of an animal's left side.

Bloat hazard is greatly reduced when at least 50 percent of a pasture is grass. Providing ready access to salt and water when grazing a legume pasture is also helpful. Bloat potential is greatest during rapid growth periods in spring, declining during the summer months. In most of the South, the period between mid-March and mid-May is when bloat is most likely to occur.

An effective method of bloat prevention on legume pastures is providing salt-molasses blocks containing surfactants, a detergent-type compound that reduces development of stable foam in the rumen. Although surfactant blocks are relatively expensive, they can prevent animal deaths. Generally, the need for surfactant blocks is greatest in early spring and they may not be needed as the grazing season progresses. Feeding an ionophore (several types are commercially available) can also reduce the potential for bloat.

Nitrate Poisoning

Cause. Nitrate poisoning in livestock is primarily caused by the consumption of pasture or hay containing high levels of nitrate-nitrogen (NO_3-N.).

During periods of low soil moisture or low humidity, nitrates can accumulate in plants heavily fertilized with N. Hay cut during or just after a drought period is suspect, especially if N was applied just prior to hay harvest. Shading by other plant species, cloudy weather, and frost may also increase nitrate levels in plants. The application of N fertilizers during cool, wet, cloudy weather may also result in nitrate poisoning. Nitrates in hay are stable and can cause deaths months after harvest.

Some forage plants are more likely to accumulate nitrates than others. Plants known to have considerable potential for accumulation of toxic levels of nitrates are: sudangrass, sorghum-sudan hybrids, pearl millet, corn, wheat, and oats. Certain weeds may also accumulate toxic levels of nitrates and thus pose a threat, especially in hay. Examples are pigweed, smartweed, ragweed, lambsquarter, goldenrod, nightshades, bindweed, Canada thistle, and stinging nettle. In addition, the application of 2,4-D herbicide can increase nitrate levels in plants.

Nitrates present in forages may be reported by laboratories as NO_3-N or as potassium nitrate (KNO_3). Plants containing more than 1.5 percent nitrate (15,000 ppm) are considered toxic to many classes of livestock (see **Appendix 15**). The reason for the toxicity is that **nitrates** are reduced to **nitrites** in the digestive tract. Nitrites then oxidize the iron in blood hemoglobin and prevent adequate oxygen transport. Animal symptoms are labored breathing, muscle tremors, and a staggering gait after which the animal collapses, gasps for breath and dies quickly (**Figure 21.2**). The membranes of the eyes and mouth are bluish, indicating a lack of oxygen. The blood is chocolate-brown but turns a brighter red when exposed to air.

Figure 21.2. Cow affected by nitrate toxicity on sorghum-sudan pasture.

Control. Animals grazing heavily N-fertilized pastures of suspect species during drought, or wet pastures during cool cloudy weather, should be carefully watched for symptoms. Prompt medication with a 4 percent solution of methylene blue supplied intravenously using 100 cc per 1,000 lb of body weight can prevent death. However, it is rare that the problem is diagnosed quickly enough to administer this treatment. Supplemental grain feeding can reduce risk through a dilution effect.

Hay well-fertilized with N and produced during drought periods should be analyzed for nitrates in a laboratory. Hay containing up to 2,500 ppm nitrate is usually safe to feed. At levels of 2,500 to 5,000 ppm, several feeding cautions are suggested. Levels of 5,000 to 15,000 are considered dangerous and require feeding restrictions. Levels over 15,000 are considered toxic to most classes of livestock and should not be fed free choice.The danger does **not** decrease with time. Toxic hay may be ground and mixed if the nitrate-containing hay is no more than 15 percent by weight of the total ration.

Animals can tolerate low levels of nitrates, but problems quickly develop when threshold levels in the blood are exceeded. Use of large round bales or stacks increases the danger of nitrate toxicity because the animals are likely to have the opportunity to consume more hay and thus get a larger total amount of nitrates in their bodies within a short period of time. Also, large bales and stacks may have concentrated spots of nitrates. When nitrate poisoning occurs, it often kills many animals in a herd because a large group of animals often have the same opportunity to gorge themselves on high nitrate forage.

Prussic Acid

Cause. Naturally occurring glycosides may form prussic acid, also called hydrocyanic acid or HCN, which can build up to toxic levels in leaves of a number of plants including Johnsongrass, sorghum, sudangrass, sorghum-sudan hybrids, and wild cherry. Pearl millet does not produce prussic acid. Prussic acid is most likely to build up to dangerous levels immediately after a killing frost. Also, tender young growth occurring immediately after a long drought can be potentially toxic. Young, tender fast-growing plants are more likely to be toxic than older, more mature plants. Herbicides, including 2,4-D, may temporarily increase prussic acid levels.

Prussic acid causes death by interfering with the oxygen-transferring ability of the red blood cells, causing animals to suffocate. Symptoms include excessive salivation, rapid breathing, and muscle spasms, and may occur within 10 to 15 minutes after the animal consumes prussic acid-containing forage. Animals may stagger, collapse, and eventually die.

Prussic acid and nitrate poisoning are **not** the same. Toxic levels of nitrates result from heavy N fertilization followed by severe drought stress. Unlike nitrates, prussic acid deteriorates with time. If forage having high levels of prussic acid is ensiled, it will usually be safe to feed within 3 weeks after silo fill. Hay which has dried enough to be safely baled (18 to 20 percent moisture) will not contain toxic levels of prussic acid. Standing plants killed by frost are normally safe after about one week. However, in some instances only plants in certain portions of a field are initially killed and subsequent frosts create danger spots in other areas.

Control. A producer should know which forage crops have the potential for prussic acid buildup. Grazing should be avoided until at least a week after the end of a severe drought. Once frost occurs, grazing or feeding greenchop should be avoided for at least a week after the last green material has been frosted. When grazing potentially toxic pastures, it may be advisable to first turn one or two low-value animals into the pasture and observe them closely. Turning hungry cattle into potentially toxic pastures should be avoided.

Grass Tetany

Cause. Grass tetany is associated with low levels of magnesium (Mg) in the blood of cattle and sheep grazing ryegrass, small grains, and cool season perennial grasses in late winter and early spring. It is mostly confined to cows and ewes in the early stages of lactation and often affects the highest-producing animals in a herd or flock. It results from animals grazing plants grown on soils low in available Mg, causing them to be deficient in this element, especially when lactation requires a substantial quantity of Mg.

Wet soils, low in oxygen, may prevent plants from taking up sufficient Mg regardless of the soil Mg level. Grass tetany is more likely to occur on soils low in phosphorus (P) but high in potassium (K) and N because this combination tends to inhibit Mg uptake. This can be a problem with cool season grass forage fertilized with high rates of broiler litter. Generally, forage containing 0.2 percent Mg or more is unlikely to cause tetany.

An animal going into tetany initially is nervous, with muscle twitching, staggers when walking, and later goes down on its side, with muscle spasms and convulsions. If not treated, death will occur.

Control. Pastures deficient in Mg should be limed with dolomitic limestone which contains this element. However, Mg fertilization cannot be expected to always prevent tetany as grass plants cannot take up sufficient Mg on waterlogged soils common in late winter and early spring. Phosphorus fertilization may also be helpful on some soils. Legumes are high in Mg, and pastures containing sufficient legume forage will normally offset the problem. However, legume growth is often limited in winter so most of the early season forage may be tetany-prone grass. The most dependable control is supplemental feeding of Mg-fortified mineral mix during the potentially dangerous tetany season. In most of the South, grass tetany is most likely to occur during the period from mid-February to mid-April on pastures of tall fescue or small grains.

Other Disorders

Acute bovine pulmonary emphysema (ABPE) is a respiratory disorder which may occur when cattle are abruptly shifted from dry feed to lush pasture or from pastures of grass to those with a high percentage of legumes. Cows that have recently calved are more susceptible. The disorder is characterized by acute respiratory distress, labored breathing, frothing at the mouth and grunting while exhaling. Occasional coughing and a frothy nasal discharge have been observed. A slight elevation in body temperature may be present. ABPE can be fatal and death can occur within 12 hours of onset of symptoms. The best prevention is to avoid *sudden* changes in diet. If an outbreak occurs, quickly remove animals from pasture and feed grass hay.

Bermudagrass staggers (bermudagrass tremors) is a nervous disorder of cattle caused by alkaloids from fungal infection of bermudagrass. Outbreaks of the disorder are rare. The problem may occur when cattle are grazing tall, mature bermudagrass pastures during autumn and winter following a period of cloudy, damp weather which promotes the growth of toxin-producing fungi. Hay cut from these pastures has remained toxic for up to two years.

Symptoms are similar to ergot poisoning. Infected cattle twitch, tremble, become stiff-legged in the hind quarter, weak in the front legs, and are poorly coordinated. Cattle may fall to their knees when excited. Animals usually recover rapidly when removed from toxic pasture and given an alternative feed source. This problem can be prevented by keeping pastures in a young vegetative stage and not grazing tall, mature, matted bermudagrass.

Ergot poisoning is caused by a parasitic fungus that grows in the seed heads of small grains, ryegrass, bahiagrass, and especially dallisgrass. A toxin produced by the ergot interferes with circulation, resulting in reduced blood flow to extremities of the limbs and tail. Lameness is often an early symptom with sloughing of the tail tip and possibly feet if cattle continue to be exposed to ergot. Other symptoms include stimulation and depression of the central nervous system, elevated body temperature, and increased respiratory or pulse rate. Since ergot grows in seedheads, clipping of pastures reduces seedhead development. Cattle should be changed immediately to an ergot-free diet. ■

Chapter 22

Poisonous Plants

MANY PLANTS which grow in the South contain compounds which can be harmful to livestock. But poisoning normally does not occur as long as the animals have access to adequate amounts of good quality forage. When pastures are short and animals are hungry, they will often eat plants they would otherwise avoid. In addition, some plants may become attractive to livestock during certain stages of growth, after a herbicide application, or when animals have access to wilted leaves of plants which have been cut or blown down.

Space does not allow discussion of all Southern plants which may be toxic. However, this chapter provides basic information regarding a few of the plants most frequently responsible for livestock poisoning within the region. Some plants discussed can also be toxic to humans. Symptoms of poisoning listed are not all-inclusive and may differ in different types of animals. Zones referred to under "Adaptation" are identified in **Figure 3.2.**

Courtesy of Dr. Tom Powe

Black Locust
(*Robinia pseudoacacia*)

Description: Small, thorny, leguminous tree with dark-colored rough bark and white flowers. Often, trees form a "thicket." Individual trees rarely exceed a height of 50 feet.

Adaptation: Found throughout the South, especially in Zones B, C and D.

Poisonous Principle: Robinin (a phytotoxin) and robitin (a glycoside).

Symptoms of Poisoning: Weakness, nervousness, nausea, vomiting, diarrhea, bloody feces, posterior paralysis, and pounding heartbeat.

Other: All of the above-ground portions of the plant can cause poisoning, especially during the warmer months of the year. Horses are more sensitive than cattle, and the most commonly encountered problem results from horses chewing the bark of trees. Usually not fatal.

Courtesy of Dr. Tom Powe

Bracken Fern
(*Pteridium aquilinum*)

Description: Herbaceous fern, usually no more than 2 to 3 feet tall, and having stout, dark-colored underground roots. Spreads primarily by underground roots.

Adaptation: Widely adapted in the South and can be found in Zones A, B, C and D. Most common in the Coastal Plain or dry or gravelly upland areas in the Piedmont or upper South.

Poisonous Principle: In monogastric animals, thiaminase (an enzyme) and possibly unidentified compounds. In ruminants, prolonged ingestion produces bone marrow suppression, possibly due to unidentified toxins.

Symptoms of Poisoning: In monogastric animals, weight loss, lethargy, convulsions, death. In ruminants, bloody feces, difficult breathing, failure of blood to clot, bleeding from all body openings, elevated temperature, death.

Other: The entire plant is toxic, either in green or dried form, but especially young green material. Death is more likely in cattle or horses than in sheep. Poisoning is most likely in late summer or autumn in years or situations in which little other feed is available. Symptoms may not appear for several weeks after consumption.

Courtesy of Dr. Tom Powe

Buckeye or Horsechestnut
(*Aesculus* species)

Description: Trees or shrubs, most having red, yellow, or yellow-green to white flowers and large (approximately 1 inch diameter) brown seeds or nuts.

Adaptation: Various species are adapted throughout Zones A, B, C and D. Most commonly found along streams or the edges of woods.

Poisonous Principle: Aesculin (a glycoside) and possibly other unidentified agents.

Symptoms of Poisoning: Weakness, muscle spasms, vomiting, dilated pupils, paralysis.

Other: All types of livestock may be poisoned by eating young sprouts or tender leaves in the spring, or by consuming seed in autumn or early winter. Not likely to be fatal.

Castorbean or Mole Bean
(*Ricinus communis*)

Description: Upright herbaceous summer annual with large leaves and reddish or purplish stems, sometimes reaching 12 feet in height.

Adaptation: Can be grown throughout Zones A, B, C, and D. Usually planted as an ornamental, but occasionally escapes and volunteers in old fields, waste places, or along roadsides.

Poisonous Principle: Ricin (a phytotoxin).

Symptoms of Poisoning: Vomiting, diarrhea, trembling, sweating, convulsions, and death.

Other: Both leaves and seeds can cause poisoning of all classes of livestock, but horses are particularly sensitive. Ricin is a deadly poison. Poisoning most frequently occurs as a result of grains or ground feed becoming contaminated with castorbean seed.

Courtesy of Dr. Tom Powe

Jimsonweed or Thornapple
(*Datura stramonium*)

Description: Herbaceous summer annual 3 to 5 feet tall with white or purplish flowers. The fruit is a spiny capsule 1 to 2 inches long.

Adaptation: Zones A, B, C, and D. A common weed in waste places, barnyards, disturbed areas, and cultivated fields, especially in the Coastal Plain.

Poisonous Principle: Hyoscyamine, atropine, and possibly other alkaloids.

Symptoms of Poisoning: Vomiting, dilated pupils, slow respiration, intense thirst, frequent urination, convulsions, and death.

Other: All parts of the plant are toxic, but especially the seeds. Dried plants mixed in hay can also be toxic. All types of livestock can be poisoned, but most cases occur with cattle. Poisoning by this plant is fairly uncommon, but can occur in summer or autumn when little other forage is available.

Lantana (*Lantana camara*)

Description: Herbaceous perennial shrub which may reach 5 feet in height. Flower color varies greatly.

Adaptation: Occurs wild in South Florida (lower portion of Zone A), but commonly grown as an ornamental throughout the South. Occasionally escapes to roadsides and waste areas or may be present around old houses.

Poisonous Principle: Lantanin (a triterpenoid).

Symptoms of Poisoning: Vomiting, bloody and watery diarrhea, lesions around the mouth and nose, weight loss, death. Also causes photosensitization (tissue death and/or lesions), especially in light colored animals.

Other: All parts of the plant are poisonous, but especially the fruit. All types of grazing livestock can be affected. Usually not fatal.

Courtesy of Dr. Ron Shumack

Mountain Laurel, Ground Ivy, Bush Ivy (*Kalmia latifolia*)

Description: Evergreen woody shrub or small tree with showy white, pinkish, or purplish flowers.

Adaptation: Usually in heavily wooded areas (often around spring heads and small streams) in the mountains or other woodlands in the upper part of Zone B, and in Zones C and D.

Poisonous Principle: Andromedotoxin (a resinoid) and possibly other toxic compounds.

Symptoms of Poisoning: Excessive salivation, frequent swallowing, vomiting, frequent defecation, staggering, coma, death.

Other: All kinds of grazing livestock may be poisoned, but sheep seem to be particularly sensitive. Poisoning usually occurs in late autumn, winter, or early spring when little else is available for animals to eat. Otherwise, animals normally avoid the plant. Several related species in the genera *Rhododendron*, *Leucothoe*, and *Lyonia* may also be toxic. Goats are especially fond of ornamental azaleas.

Courtesy of Dr. Tom Powe

Courtesy of Dr. Ron Shumack

Nightshades (*Solanum* species)

Description: Annual or perennial herbs, some species covered with spines.

Adaptation: Various species are common in pastures, waste areas, or cultivated fields throughout Zones A, B, C, and D.

Poisonous Principle: Solanine and other alkaloids.

Symptoms of Poisoning: Excessive salivation, drowsiness, trembling, vomiting, diarrhea, dilation of pupils, paralysis, death.

Other: Poisonous or potentially poisonous species include deadly nightshade (*Solanum americanum*), horsenettle or bullnettle (*Solanum carolinense*), and black nightshade (*Solanum nigrum*). The leaves and especially the unripe fruit (berries) are poisonous to cattle, horses, and sheep. Apparently, numerous environmental factors, plant maturity, and possibly animal breeding influence susceptibility. In some cases animals have been known to selectively graze nightshades.

Oaks (*Quercus* species)

Description: Trees or woody shrubs, usually deciduous, the fruit or nut being an"acorn."

Adaptation: There are many species of oaks, and there are numerous adapted species in Zones A, B, C, and D. Most species are best adapted to rich, well-drained soils.

Poisonous Principle: Tannins and possibly other unidentified compounds.

Symptoms of Poisoning: Constipation followed by bloody diarrhea, poor appetite, weakness, rough appearance, thirst and excessive urination.

Other: Poisoning of all types of livestock may result from consumption of leaves or acorns. Poisoning is unlikely unless there is little else for animals to eat, but once symptoms are observed death is likely even with treatment. Some of the more common species are white oak (*Quercus alba*), Southern red oak (*Quercus falcata*), black oak (*Quercus velutina*), post oak (*Quercus stellata*), water oak (*Quercus nigra*), and blackjack oak (*Quercus marilandica*).

Courtesy of Dr. Tom Powe

Oleander (*Nerium oleander*)

Description: Evergreen shrub or small tree with leathery leaves and reaching a height of up to 25 feet. Flowers white, yellow, pink, or red in clusters on stem tips.

Adaptation: Used as an ornamental primarily in the Coastal Plain (Zone A). May also be found around old buildings or growing wild in fencerows, along the edges of woods, or in waste places in some areas along the coast.

Poisonous Principle: Various glycosides and possibly other toxic compounds.

Symptoms of Poisoning: Vomiting, trembling, weakness, bloody diarrhea, death.

Other: Leaves and twigs can cause poisoning of all types of livestock. Poisoning usually occurs when animals have access to clippings. Dried or green foliage or other plant parts can be highly toxic. Smoke from burning plants can produce toxic symptoms.

Courtesy of Dr. Tom Powe

Perilla Mint (*Perilla frutescens*)

Description: Herbaceous annual with small whitish or purplish flowers and usually no more than 1 to 2 feet tall.

Adaptation: Common throughout Zones A, B, C, and D in pastures, waste places, and along roadsides.

Poisonous Principle: Unknown.

Symptoms of Poisoning: Difficult breathing, nasal discharge, death.

Other: Apparently, all aboveground portions of the plant can be toxic. Poisoning is most common in cattle or horses, usually in late summer or autumn.

Courtesy of Dr. Tom Powe

Poison Hemlock
(*Conium maculatum*)

Description: Herbaceous warm season biennial resembling wild carrot. Stems hollow except at nodes. May reach a height of 6 to 8 feet.

Adaptation: May be found in wet-natured waste areas or ditches throughout Zones C and D.

Poisonous Principle: Several alkaloids, especially coniine and N-methyl coniine.

Symptoms of Poisoning: Bloody feces, vomiting, convulsions, paralysis, death.

Other: All portions of the plant are toxic. Toxicity increases as the plant matures, but decreases as it dries after being cut or frosted. If poisoning occurs, it is usually in the autumn. Grazing animals generally avoid it unless little other feed or forage is available.

Courtesy of Dr. Tom Powe

Pokeweed (*Phytolacca americana*)

Description: Upright perennial herb, reaching a height of up to 10 feet.

Adaptation: Common throughout Zones A, B, C, and D in waste areas, along roadsides, or adjacent to wooded areas.

Poisonous Principal: Phytolaccotoxin (a saponin) and possibly other unidentified compounds.

Symptoms of Poisoning: Vomiting, convulsions, death.

Other: The leaves, berries, and especially the roots are toxic. Cattle and occasionally horses are poisoned, but poisoning is fairly rare. Poisoning most commonly occurs during hot, dry periods in summer or autumn. Cattle have reportedly aborted after eating this plant.

Courtesy of Dr. Ron Shumack

Rhododendron

(*Rhododendron* species)

Description: Woody evergreen or deciduous shrub or small tree up to 15 feet tall. Flowers showy and may be white, rose, or pink.

Adaptation: Grows mainly at higher elevations in the mountains or the upper Piedmont in Zones C and D. Usually found in moist soils in heavily wooded areas.

Poisonous Principal: May contain at least two toxic compounds: andromedotoxin (a resinoid) and arbutin (a glucoside).

Symptoms of Poisoning: May include nausea, vomiting, incoordination, paralysis, and death.

Other: Flowers, leaves, and young stems may be toxic to cattle, sheep, or goats. Poisoning from plants in pastured areas usually occurs during winter or spring. However, due to the ornamental nature of the plants, trimmings or cuttings used for decorations can also be a source of toxicity. Related species in the genera *Kalmia*, *Leucothoe*, and *Lyonia* may also be toxic.

Courtesy of Dr. Mike Patterson

Showy Crotalaria

(*Crotalaria spectabilis*)

Description: Herbaceous summer annual legume with bright yellow flowers, usually no more than 2 to 3 feet tall.

Adaptation: Widely adapted in the South, but most common in the Coastal Plain or dry upland areas in the Piedmont (Zones A and B).

Poisonous Principle: Monocrotaline (an alkaloid).

Symptoms of Poisoning: May be acute or chronic. Excessive salivation, nasal discharges, bloody feces, lethargy, weight loss, aimless walking, head pressing, death.

Other: All types of livestock can be poisoned by any portion of the plant. The seed are especially toxic. If mixed in hay, the dried plants and/or seed can cause poisoning months after baling. The onset of symptoms may occur several weeks to months after consumption. This plant produces very hard seed, and volunteer stands can occur when soil is disturbed after many years with no evidence of the plant. Several other species in the genus are also believed or known to be toxic.

Wild Cherry or Wild Black Cherry (*Prunus serotina*)

Description: Woody shrubs or trees, usually small, with white or pink flowers. The fruit usually has a fleshy outer layer with a single hard stone in the center.

Adaptation: Commonly found in fencerows, borders of woods, and in waste places throughout Zones A, B, C, and D.

Poisonous Principle: Hydrocyanic (prussic) acid.

Symptoms of Poisoning: Labored breathing, spasms, rapid death. Low level poisoning may cause abortions.

Other: All types of grazing animals are susceptible. Many environmental factors may influence hydrocyanic acid levels, including drought and frost. However, most poisoning occurs when animals gain access to the wilted leaves of plants recently cut or blown down. There are many related species in this genus which are also presumed potentially toxic, but wild black cherry is considered the most dangerous.

Yellow Jessamine or Carolina Jessamine (*Gelsemium* species)

Description: Perennial trailing woody vine with fragrant yellow flowers, vegetatively resembling honeysuckle.

Adaptation: Common in the Coastal Plain and Piedmont (Zones A and B). Usually found in thickets, along the edges of woods or stream banks, or in fencerows.

Poisonous Principle: Gelsemine, gelseminine, and gelsemoidine (alkaloids).

Symptoms of Poisoning: Weakness, slow breathing, dilation of pupils, convulsions, and death.

Other: All portions of the plant are toxic to all types of livestock. Poisoning normally occurs only in winter or early spring in situations in which little other forage is available and when large quantities are consumed.

Courtesy of Dr. Tom Powe

Yew (*Taxus* species)

Description: Evergreen shrubs or (usually) small trees having needle-type leaves, reddish-brown bark, and reddish or scarlet berries.

Adaptation: Found primarily at higher elevations in Zones C and D. Also commonly used as an ornamental.

Poisonous Principle: Taxine (an alkaloid).

Symptoms of Poisoning: Staggering, muscular incoordination, death.

Other: The bark, leaves, and seeds are poisonous to all types of livestock. Highly toxic but rarely eaten. Most poisoning occurs when cuttings are thrown into pastures or when plants in pastures are cut or blown down.

Summary

Although numerous plants commonly found in the South can be toxic, including many not mentioned in the preceding discussion, poisoning is fairly uncommon. To reduce the likelihood of livestock losses from poisoning, the following steps can be taken:

(1) Learn to recognize common poisonous plants and eliminate them from areas to which livestock have access.

(2) Do not allow livestock to have access to freshly cut plants, especially of species known to be potentially toxic.

(3) Use fences to exclude livestock from unimproved areas which may contain poisonous plants.

(4) Provide adequate amounts of improved pasture or stored feed to livestock at all times.

(5) Immediately call a veterinarian if poisoning is suspected. ■

Chapter 23

Fescue Toxicity

TALL FESCUE, grown on more than 35 million acres, is the most important cultivated pasture grass in the USA. Since the late 1930s, this cool season perennial has been widely planted in an area from the lower Midwest to eastern Oklahoma and central Georgia. Its rapid acceptance was due to many attributes, including ease of establishment, wide adaptation, long grazing season, tolerance to environmental stress, and pest resistance. Over 8.5 million beef cows and nearly 700,000 horses are maintained on tall fescue pastures in the USA.

Chemical measures of forage quality such as digestible dry matter, crude protein, and mineral levels suggest that tall fescue forage has potential for high quality and should give good animal performance. However, tall fescue gained a reputation of toxicosis syndromes which have been manifested in cow reproduction problems, low calf weaning weights, and poor gains of growing steers. Tail switches and hooves sometimes develop a gangrenous condition and fall off, and animals often have scruffy hair coats. Mares grazing tall fescue often abort, produce dead foals, have foaling difficulties, or fail to produce enough milk.

The Fescue Toxicosis Syndromes in Cattle

There appear to be three separate syndromes associated with tall fescue. A brief description of each follows.

1. Fescue foot. The clinical signs are rough hair coat, loss of weight, elevated body temperature and respiration rate, tenderness of legs, and loss of hooves and/or tail switch. Fescue foot occurs mainly in winter and is mostly confined to the upper South. It is a serious syndrome but its occurrence is relatively low in relation to the large acreage of tall fescue.

2. Bovine fat necrosis. It is characterized by the presence of hard masses of fat in the adipose tissue, primarily in the abdominal cavity. Necrotic fat lesions are most common in the fat surrounding the intestinal tract all the way to the rectum. This syndrome, which results in upset digestion and difficult births, has been associated with high rates of nitrogen (N) fertilization from either chemical fertilizers or from poultry litter commonly applied at heavy rates in the poultry-producing areas of northern Arkansas, Alabama, and Georgia.

3. Fescue toxicity or summer slump. It is characterized by poor animal gains, reduced conception rates, intolerance to heat, failure to shed the winter hair coat, elevated body temperature, and nervousness. This syndrome occurs throughout all tall fescue-growing regions of the USA. The term "summer slump" has been applied in the northern tall fescue-growing regions where it is most noticeable in summer, but the adverse effects occur throughout the year. Fescue toxicity is by far the most common syndrome in cattle on tall fescue.

Fescue Toxicity Syndrome in Horses

Fescue toxicity causes serious reproduction problems with mares. Specific problems are abortions, prolonged gestation, dystocia (difficult birth), thick placenta, foal deaths, retained placentas, agalactia (little or no milk production), and sometimes death of mares during foaling.

Cause of Fescue Toxicity

For many years, various alkaloids occurring in tall fescue were suspected to cause the syndrome but this was never proven. In 1977, several scientists at the USDA Russell Research Center, Athens, Georgia, reported a fungus in the grass that was suspected to be associated with the syndrome. This was confirmed in grazing trials in Alabama where excellent animal gains and no evidence of toxicity were obtained on fungus-free (endophyte-free) tall fescue. The fungus associated with the toxicity is an endophyte (lives within the plant) and is called *Acremonium coenophialum.* The fungal endophyte produces ergot alkaloids, such as ergovaline, which can be highly toxic to livestock.

Animal Response

Beef steer daily gains on endophyte-free tall fescue are typically 50 to 100 percent more than on heavily infected grass. Steer daily gains are 1.5 to over 2 lb per day on endophyte-free grass, compared to around 1 lb or less on heavily infected grass. Steers on endophyte-free grass are tolerant of heat, graze throughout the day, shed their winter hair coats in spring, and usually are less nervous than steers on endophyte-infected grass (**Figure 23.1**).

Visible signs of the syndrome increase with higher air temperatures, but poor gains occur throughout the year on endophyte-infected tall fescue. Cattle with

Figure 23.1. Beef steers grazed from November to May on infected (left) and endophyte-free (right) tall fescue in the Black Belt of Alabama. Steers grazed on endophyte-free tall fescue gained more weight and had a slick hair coat.

Figure 23.2. Beef cows and calves maintained on endophyte-infected tall fescue pasture were thin and retained their winter hair coat even in May (Black Belt of Alabama).

Brahman breeding are more heat-tolerant and may exhibit fewer symptoms of fescue toxicosis. Nonetheless, their gains are adversely affected.

Beef cows on endophyte-infected tall fescue are generally thin and in poor condition (**Figure 23.2**). Conception rate of beef cows may also be reduced, often to 50 percent or less. Calf weight gains are decreased, a result of both reduced milk production by cows and consumption of toxic tall fescue forage by the calf. Milk production on endophyte-infected grass can be reduced by as much as 50 percent.

Lactation of dairy cattle is also reduced on endophyte-infected tall fescue. However, milk production of dairy cows grazing endophyte-free fescue has been equal to that on annual ryegrass.

Horses are quite sensitive to infected tall fescue (**Figure 23.3**). Mares grazing the grass may abort or have dead foals and produce little or no milk. The evidence is overwhelming that these problems are associated with the presence of the endophyte.

Endophyte-infected tall fescue hay and seed also contain the toxic substance, and can result in poor animal performance. Hay from endophyte-infected fields has remained toxic even when stored for two years.

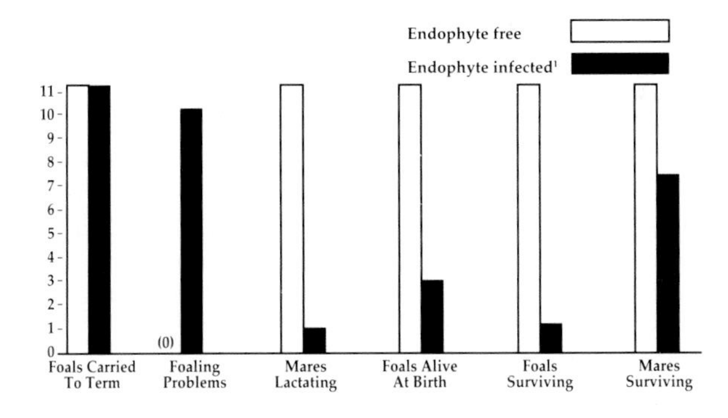

Figure 23.3. Effects of the fungal endophyte on pregnant mares and foals grazing tall fescue.

[1]Greater than 80% of plants testing positive for *A. coenophialum* in analyses conducted in the Auburn University Fescue Diagnostic Laboratory.

Source: M.R. Putnam, D.I. Bransby, J. Schumacher, T.A. Boosinger, L. Bush, R.A. Shelby, J.T. Vaughn, D.M. Ball, and J.T. Brendemuehl. 1991. Am. J. Vet. Res., Vol. 52.

Biology of the Fungal Endophyte

The fungus lives its entire life cycle within the plant, thus being called an endophyte. Unlike most fungi, it is not visible externally on the tall fescue plant. Infected grass appears the same as endophyte-free tall fescue, and the only way the presence of the fungus can be detected is by laboratory examination.

The endophyte does not appear harmful to the tall fescue plant. In fact, the endophyte and the grass derive mutual positive benefits from their association. Endophyte-infected tall fescue is more tolerant of drought, nematodes, and insects.

All evidence indicates that the endophyte is spread only through infected seed. This is important because it means that tall fescue pastures free of the endophyte will remain that way for a long time. Invasion of an endophyte-free pasture could occur if infected seed are introduced, but at some locations endophyte-free tall fescue pastures adjacent to heavily infected pastures have remained free of the endophyte for several decades. However, pastures having a low or intermediate level of infection can be expected to slowly increase in percentage of infected plants, probably due to the higher stress tolerance (competitive advantage) of infected plants.

Storage of infected seed under normal conditions generally results in death of the endophyte after 1 to 2 years. Thus, planting old seed of endophyte-infected tall fescue can result in endophyte-free pasture. However, seed germination and vigor also decline with storage time. Planting of vigorous, new crop, low-endophyte seed from clean seed fields is the best way to obtain an endophyte-free pasture.

Solutions to the Fescue Toxicity Problem

Surveys in several states indicate that most tall fescue pastures are endophyte-infected at levels averaging 60 percent to over 80 percent. The widespread infection can be attributed to the seed increase and subsequent planting of fresh Kentucky 31 fescue seed that originated from one heavily endophyte-infected seed source in Kentucky.

Not all tall fescue pastures are infected with the endophyte, so it is important to determine the level of infection. If a serious animal toxicity problem exists, the answer is obvious. If there is a question about infection level, each field should be tested. Several states have laboratories for determining infection levels. Agricultural Extension agents can provide information on cost, sampling methods, and laboratory addresses.

Practical solutions to the fescue toxicity problem depend on the type of livestock operation, goal expectations, and management expertise. When grazing young beef cattle, it is imperative to destroy heavily-infected tall fescue pastures and replant with endophyte-free seed. Growing steers and heifers are highly sensitive to fescue toxicity, with daily liveweight gains being reduced by about 0.1 lb for each 10 percent of fungus infection. Likewise, dairy and horse producers should consider replanting infected pastures since performance may be affected even at low levels of endophyte infection.

The question as to whether to replant infected pastures is less simple for beef cow-calf producers. If tall fescue pastures on a farm have a record of cows with poor pregnancy rates, low calf weaning weights, and visible toxicity symptoms, then replanting of pastures with endophyte-free seed will improve animal performance and profitability.

Replanting infected pastures with endophyte-free seed requires considerable capital outlay, takes pastures out of production for a year, and requires good establishment techniques. Also, better grazing management is required to maintain stands of endophyte-free pasture. This may not be attractive to cow-calf producers where fescue toxicity problems are less pronounced or where low-input is desired. Several other management options can be used to reduce endophyte effects when grazing infected tall fescue pastures.

1. Management of pastures to maintain other grasses such as Kentucky bluegrass, orchardgrass, or common bermudagrass in the sward to dilute the toxic tall fescue. Early and close grazing of tall fescue in spring prevents shading of other grass species. Summer application of N favors summer-growing grasses such as bermudagrass.

2. Close grazing of tall fescue in spring will reduce seedhead production. Mowing of seedheads will also reduce intake of the highly toxic seed by livestock. In addition, close grazing of infected fescue has reduced endophyte effects on unheaded tall fescue.

3. Seeding of legumes such as alfalfa, birdsfoot trefoil, red or white clover into pastures will greatly dilute the toxic fescue and often overcome any problems with animal gains.

4. Feeding of hay other than toxic tall fescue, such as endophyte-free tall fescue, bermudagrass, orchardgrass, timothy, alfalfa, red clover, or sericea lespedeza greatly reduces the toxicity problem in winter. Feeding grain has been beneficial to cattle consuming toxic tall fescue but has not been of benefit to pregnant mares.

Replanting Infected Tall Fescue with Endophyte-Free Seed

To successfully establish low-endophyte tall fescue pastures, the old sod must be **completely** killed. On land suitable for cropping, tillage and cropping for a year before replanting tall fescue are effective. However, many tall fescue pastures are too hilly for tillage and severe soil erosion would result if they were tilled. In addition, tillage is expensive. An attractive alternative is killing the sod with herbicide and drilling low-endophyte seed with a sod-seeder.

A good method is to kill the sod in spring with herbicide, grow a summer annual grass, and then apply herbicide again in autumn before replanting endophyte-free seed. In the year preceding replanting, the old infected tall fescue should not be allowed to make seed. Otherwise, the field will contain infected seed which may result in volunteer infected plants.

Management of Low-Endophyte Tall Fescue

Endophyte effects on stress tolerance of tall fescue have important implications for livestock producers. More precision is required in establishment of endophyte-free tall fescue pastures. This includes careful attention to soil fertility, as well as planting date, rate, and depth. It may be difficult to establish and maintain endophyte-free tall fescue pastures in areas where tall fescue is only marginally adapted.

Figure 23.4. Endophyte-free Kentucky 31 tall fescue harvested every three weeks all year had a poor stand when cut at 1.5-inch stubble (left), but a good stand when cut at 3-inch stubble (right). Third year. Athens, GA.

Low-endophyte tall fescue stands can be weakened by overgrazing and drought to a greater extent than infected tall fescue. The high palatability of endophyte-free tall fescue may contribute to stand losses due to overgrazing. The endophyte also favors greater drought tolerance of tall fescue. It accomplishes this through several mechanisms, including improved root growth and greater sugar accummulation for survival and regrowth. Thus, endophyte-free tall fescue pastures will require better grazing management to maintain stands and productivity.

Low-endophyte tall fescue pastures, planted in autumn, should not be grazed until plants are well established in late spring. Close grazing of the newly established pasture in summer weakens plants and reduces production the next autumn. In the southern area of tall fescue adaptation, environmental stresses are greater than farther north, making it even more important to avoid overgrazing endophyte-free tall fescue (**Figure 23.4**).

Summary

Fescue toxicity in livestock is caused by a fungal endophyte within tall fescue plants which produces toxic compounds. Most tall fescue pastures are infected at levels of 60 to over 80 percent, resulting in widespread losses. Annual losses from reduced beef cow conception rates and calf weaning weights caused by fescue toxicity are estimated at over $600 million in the USA. Serious losses are also encountered in horses from abortions and foal deaths.

Solutions to the problem are destruction of endophyte-infected pastures and replanting with endophyte-free seed, diluting infected pastures with other grasses or legumes, close grazing in spring, and feeding hay other than toxic tall fescue. A higher level of management is required for establishment and maintenance of endophyte-free than infected tall fescue stands. ■

Chapter 24

Fencing

LIVESTOCK PRODUCERS realize the necessity and importance of fencing. However, fencing is usually viewed as a tool for **animal** management. Typically, it is seen as being useful primarily for the purposes of keeping animals on the property, out of crop fields, away from roads or other dangerous areas, and as a means of keeping certain animals separated.

Although fencing is important for these and other animal management reasons, it is also necessary for good **forage** management. The importance of fencing as a forage management tool is sometimes overlooked; its potential is frequently not fully appreciated; and fencing configurations often contribute to inefficient animal production and/or forage utilization.

Use of Fencing in Forage Management

Some of the ways fencing can be used in forage management are readily apparent and are regularly practiced in the South. Most livestock producers use fencing to keep animals out of newly-planted pastures which would be damaged by grazing too early. Fences are also commonly used to periodically exclude livestock from fields containing stands of certain forage crops which can be damaged by overgrazing. In addition, fences are used to keep animals out of pastures sprayed with herbicides which have grazing restrictions; they protect hayfields where livestock would trample and waste accumulated forage; and they restrict the access of animals to

Figure 24.1. Fences can be used to regulate the amount of forage to which animals have access, as shown in this scene near Athens, Georgia.

fields where forage is "stockpiled" for later grazing.

Fencing can also be used to implement various grazing systems (see Chapter 25). In situations where a high level of animal output per acre is the goal, rotational stocking might be used. If maximum use of the forage available from a single undivided field is desired, limit grazing can be employed. If increasing gains for calves is a major objective, creep grazing is a possibility. These and other grazing techniques require proper fencing.

Many livestock producers could use additional fencing to do a better job of concentrating their animals on a smaller acreage during periods of lush forage growth, thus helping keep forage quality high in the grazed area while allowing excess growth in other areas to be saved for hay or silage.

A primary reason for the efficiency of livestock production in countries such as New Zealand is the use of grazing methods which allow efficient use of the available forage. This efficiency does not result from the use of a single grazing method, but rather from flexibility in altering grazing management to fit the present situation and objectives. Fencing is an integral part of efficiently and effectively implementing grazing methods.

Basic Fencing Decisions

Planning the optimum location and type of fencing for a livestock farm is an important step. However, the expense of constructing new fences makes it important to make the right decisions. When existing fences are acceptable, they should be used. The useful life of an old fence can often be extended by adding electric wires at the top and bottom.

Cattle producers are generally best served by erecting barbed wire or high tensile wire (HTW) fences or some combination of the two. Electric fences were once considered to be "temporary," but that is no longer true. Although some producers use HTW fences exclusively, others feel more comfortable with barbed wire fences around the perimeter of their property and HTW fences in the interior. Regardless of the type of fence built, accurate determination of external property boundaries and dependable restriction of livestock therein is of critical importance.

Barbed wire is usually shunned by horse producers because of the possibility of injury. Sheep producers avoid barbed wire because of the possibility of wool entanglement, and appreciate the predator control features of HTW fencing. Therefore, an increasing number of horse and sheep producers are using HTW fences. Horse producers may also use board fences, but the expense of such fences can be prohibitive. Because of the availability of HTW fencing, the use of costly woven wire fences is declining.

When making fencing decisions, the objectives should constantly be kept in mind. Most producers are primarily interested in building a fence which will effectively keep livestock in the pasture, be easy to erect, and be as inexpensive as possible. Other concerns might be: keeping dogs, coyotes, and other animals (or humans) **out** of the pasture; ease of repair; desired life of the fence; and attractiveness.

Determination of pasture size and the exact placement of fences are highly important decisions. Large pastures are attractive, but they often are difficult to manage and inefficient. Field sizes should be kept small enough to allow for efficient grazing and cutting management of different forages to obtain high animal performance.

Pasture size should be dependent upon herd size and overall livestock numbers, but many producers in the South find that pastures of 10 to 20 acres work well. However, in some intensive grazing situations, it may be desirable to have much smaller pastures or paddocks. Pastures do not have to be the same size and shape, but it is generally simpler to standardize them if possible. Long, narrow pastures should be avoided because they require more fencing. Access to water and shade for grazing animals is usually an important constraint to pasture size and configuration.

It is necessary to separately fence areas for different forage species. Since terrain and soil type largely determine this, it is desirable to run fencing along natural dividing lines for various forage species. For example, on a farm in Zone B (see **Figure 3.2**), low-lying, relatively flat bottom land might be used for tall fescue and white clover, and an adjacent upland area might be best suited to bermudagrass or sericea lespedeza. Thus, the fence line might be placed along the "break" between the bottom and the upland area to separate them. This allows differential grazing and/or cutting management which will favor the species which is/are expected to dominate in each area.

Other concerns which must be considered are water and shade for the animals. Also, the location of catch pens and/or buildings such as hay barns may influence fence placement. Furthermore, the exact location of gates should be determined in advance of fence construction. The number and length of lanes should be minimized since they increase fencing costs. When possible, fences should be located away from wooded areas in which trees may fall on the fence.

Because of the need to carefully plan each aspect of fencing, it is best to sketch on paper the optimum location of fences prior to fence construction. This should be done only after visually inspecting the soils and terrain, and with the benefit of an aerial or scale-drawn map. This allows an accurate estimate of the quantity of fence materials which will be needed.

There are numerous types of fences, and there is a great deal of technical information which could be provided regarding each type. Space does not allow a comprehensive discussion of fences in this book. However, some basic concepts and facts regarding barbed wire and high tensile wire fences are provided.

Barbed Wire Fences

The development and availability of barbed wire has been of major importance to the livestock industry in the USA. Generally, three strands of barbed wire, with enough posts (including large corner posts) to space them at 15 to 20 foot intervals, and enough galvanized fence staples to attach the wire to the posts are all that is needed.

Barbed wire is available in various gauges, with varying numbers of points per barb (usually 2 or 4), and various spacings of the barbs (usually 4 or 5 inches apart). Barbed wire is galvanized or aluminum coated to resist corrosion. Staples, usually 1.5 or 2 inches in length, are normally used to attach barbed wire to wooden posts.

The type of wood used for posts largely determines the life of the fence. Osage orange (also called "bodock" in some areas) makes one of the best fence posts, often lasting 30 years or more. Other good choices are black locust, red cedar, and mulberry, each of which may last up to 20 years. White oak will usually last around 10 years, but pine and other soft woods will last only 2 to 7 years.

Therefore, for woods other than osage orange, red cedar, black locust, or mulberry, it is best to treat the posts. Pressure treating with a preservative can increase the life of "soft wood" posts to 25 years or more.

Once the fencing materials have been obtained, the corner posts, gate posts, and brace posts should be firmly set. Such posts should be a minimum of 8 inches in diameter. Line posts should normally have a diameter of at least 4 inches. The strength of a post increases greatly as the diameter increases. For example, a 5-inch diameter post may be twice as strong as a 4-inch post.

The depth of the hole is also a key to post strength. For example, a post set 3.5 feet deep is far stronger than one set 2.5 feet deep. Thus, corner posts need to be longer, as well as bigger in diameter than the line posts. In sandy soil, it may be desirable to set the corner and gate posts in concrete to insure long-term stability.

Steel line posts can also be used for barbed wire fences. Steel posts have the advantage of being lighter and easier to work with; they can be driven in the ground, and they will last for many years. However, if a steel post is bent, it may be difficult to repair, or unattractive if repaired. Clips, rather than staples, are used to attach barbed wire to steel posts.

Care should be taken to properly brace each corner or gate post. A corner post needs a brace for each fence line attached to it. Brace posts should be at least 4 to 5 inches in diameter. The commonly-used "H brace," and also a "double H brace" are shown in **Figure 24.2.** Cross posts should be trimmed as necessary to tightly fit between the corner or gate post and the brace post, then securely nailed in place. A diagonal brace wire twisted to increase tension will further stabilize the brace. Fence lines should have double braces spaced at 600 to 1,000 foot intervals within the fence line, with the closer spacings being used for sandy or loose soils.

Once the corner and gate posts are set, a wire should be strung exactly where the fence will run. Then the line posts can be set using the wire as a guide to obtain a straight line. If brush or other obstructions are in the fence line, they should be removed. Line posts should be set at least 2.5 feet deep.

Spacing of the wires is important. For cattle, a top wire height of 36 to 40 inches, and a bottom wire height of 16 to 18 inches is about right. Ideally, each post should be measured and marked in order to insure an even height. The bottom wire should be stapled first and, unless concern over appearance dictates otherwise, the wire should be placed on the **inside** of the fence so that when livestock rub against it the staples will not be loosened.

A wire stretcher is needed to pull each strand of wire tight along the line of posts. Once tight, the wire should be stapled to each post and the process repeated for the other wires. Staples should be driven firmly, but not so tightly as to pinch the wire. Driving staples parallel to the grain of the wood should be

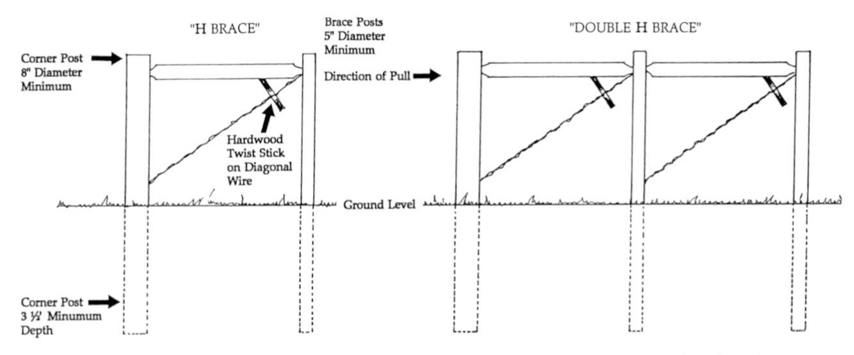

Figure 24.2 Diagram showing "H brace" and "double H brace" for fencing.

avoided, because to do so increases the likelihood of the post splitting.

High Tensile Wire (HTW) Fences

A relatively recent development, HTW fencing has the potential of helping livestock producers to more easily and economically implement improved systems of grazing. When electrified, HTW systems differ from conventional electric fences in that they have a high voltage (more than 5,000), and the pulse length is short, perhaps as short as one 10,000th of a second. The electrical resistance for these fences is lower and the fence is "hot" over a longer distance. Up to 30 miles of fence can be electrified with one energizer. This technology, developed in New Zealand, is so effective in animal control that it is used in some parts of the world to contain or exclude deer, elephants, and other wildlife.

High tensile wire fencing can be rapidly erected and quickly tightened, it is quite effective in confining animals, the smooth wire does not injure animals, and it is considerably less expensive than conventional fencing (**Figure 24.3**). Because of these advantages, it seems inevitable that the popularity of this type of fencing will continue to grow in the South in future years.

High tensile wire fences are usually electrified. Although a "permanent" HTW fence is much cheaper to erect than a barbed wire fence, it requires a wider assortment of materials.

An important difference between the old type electric fences and the new HTW fences is the energizer used to electrify the wires. The new energizers are more expensive, but, with their high voltage and short pulse they provide much more effective animal control.

The distance which an energizer can power is affected by a number of factors, including soil moisture, effectiveness of the ground, the amount of weeds and brush which are likely to come in contact with the fence, and the type(s) of animals to be confined. Sheep are less sensitive to electricity than cattle, and require a "hotter" fence. It is advisable to be on the safe side by purchasing an energizer which has a higher rating than is believed to be needed. Highly effective solar powered energizers are available for powering HTW fences in remote locations.

Figure 24.3. High tensile wire fences are rapidly gaining in popularity in the South (photo taken in central Kentucky).

Fiberglass posts are usually sold with HTW fences. However, many livestock producers use wooden posts for the corners and gates of "permanent" HTW fences. Although wooden posts can also be used for line posts, many producers use fiberglass or plastic posts because they are so easy to install. Also, long-lasting posts made from Australian ironbark trees are sold in some areas for use with HTW. The process of locating HTW line posts is identical to that for building barbed wire fences except that once the corner and gate posts are securely in place, fiberglass or plastic posts can be driven, rather than set, as line posts. The flexibility of plastic posts, which allows them to be bent over temporarily, is a feature which many producers especially like.

A major difference between conventional electric fences and HTW fences is that the latter must be insulated better. If unruly stock are to be kept confined, it is also advisable to use heavy wire offset brackets which will help keep animals away from the posts.

HTW fences are usually 11 to 14½ gauge wire and have a breaking strength of up to 1,800 lb. HTW fences should be stretched to 150 to 250 lb of tension. Heavy leather gloves and eye protection should be worn when tensioning wire fences.

It is extremely important to properly ground HTW fences. Care should be taken to install the ground rod according to instructions provided with the energizer. Most electricity "leaks" with HTW fences pertain to improper insulation and/or poor grounding.

There are many accessories available for use with HTW fences, including spacers which can be used to separate and stabilize wires, voltage meters, lights which flash to indicate the fence is working, and electric fence warning signs.

The necessary number of wire strands varies with the use of the fence. For cattle, three strands are normally used for exterior fences, while two strands should be adequate for interior fences. Use of four or five strands will hold sheep and keep out predators if properly spaced. Dogs and coyotes can be repelled by spacing of the bottom two wires at about 6 and 10 inches.

Eight to ten strands of wire can contain livestock even without being electrified. If HTW for a non-electrified fence is attached to wooden posts with staples, the staples should not be driven tight and thus allow for expansion and contraction of the wire. However, electrification of one or both of the bottom wires (requiring insulators) may be desirable to repel predators. Also, to insure the safety of animals and humans, a lightning arrestor should also be installed at various intervals along the fence.

A final important feature of HTW fences is the installation of tension springs and/or ratchet tighteners at various spots along the fence. These are necessary to compensate for the contraction and expansion of the wire as the temperature changes throughout the year. The optimum placement of tension springs varies with the length and design of the fence. Informational materials which explain this are available from HTW dealers.

Water

How a farm is fenced and the number and locations of water sources provided for livestock are directly related topics which must be considered simultaneously. Each separately-fenced area must provide access to at least one source of clean water. In many fencing configurations one source can provide water to two or more pastures, but as the number increases, trampling damage near the water source increases. In general, the most efficient fencing systems will provide pastures having low length-to-width ratios with a source of water in each one, or a central alley which provides access to water.

Location of water is quite important. The further livestock have to walk to get to water, the more energy and time they spend, and the lower their performance is likely to be. Water sources should be positioned on level ground if possible to reduce erosion hazard created by hoof action and to minimize the amount of energy the animals have to expend to reach water.

Economics

The cost of fencing a given area depends mainly on the type of fence, length, the number of wire strands, wire quality, and type and spacing of posts. Also, the costs of fencing materials vary depending on location and are subject to change. Therefore, no specific figures regarding fence costs are provided here. However, the cost per unit length of an HTW fence is usually about 50 percent as much as barbed wire and may be less than 30 percent as expensive as woven wire. ■

Chapter 25

Grazing Management

MANAGEMENT of pasture plants is unique. While most crop plants are grown for their fruit or seed, the leaves or photosynthetic area of plants in a pasture are continually or frequently being removed. This places great stress on the plants, often making them dependent on ungrazed leaves and stored food reserves in the roots, rhizomes, stolons, or stem bases to fuel regrowth after grazing or cutting (see Chapter 14).

Pasture Growth

High pasture leaf production over an extended period of time depends on a number of factors, including:

(a) Effective light interception and high photosynthetic efficiency,

(b) High sustained growth rate,

(c) Low rate of leaf aging and decomposition of leaves,

(d) Maintenance of a vigorous tiller or shoot population for regrowth.

Balancing these factors to achieve high forage productivity and animal performance may appear complex, but can be done when the dynamics of pasture growth are understood. Pasture yield potential depends on the amount of light intercepted by the leaves. Considering a leaf area unit of 1 as one layer of leaf tissue covering an equivalent land area, pasture yield potential increases as leaf area increases until leaves intercept about 90 percent of the light (**Figure 25.1**).

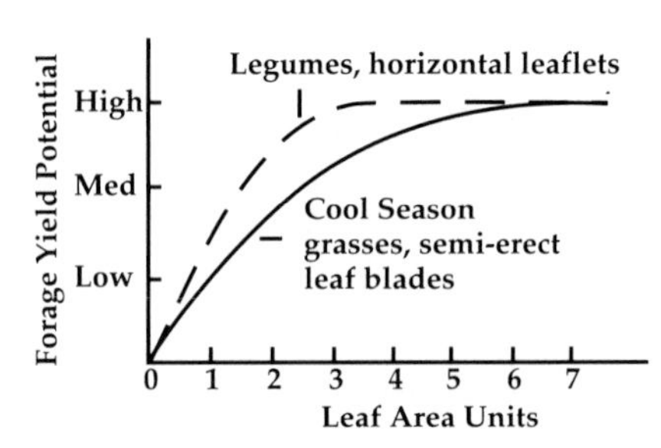

Figure 25.1. Forage yield in a pasture increases as leaf area increases until leaves intercept about 90 percent of the light.
Adapted from: R.E. Blaser. 1986. Virginia Agric. Exp. Stn. Bull. 86-7.

Cool season grasses such as tall fescue generally reach this point at about 6 leaf area units while some warm season grasses such as pearl millet with more upright growth allow more light to penetrate the sward, and may attain up to 10 or 12 leaf area units. Legumes such as white clover with horizontal leaves intercept more light per leaf and cause greater shading of other leaves, thus reaching the forage production potential plateau at about 3 leaf area units. The practical significance of this is that an overgrazed grass pasture will have a greatly reduced leaf area of perhaps only about 2 leaf area units and thus is not able to utilize sufficient light to reach its yield potential. Close grazing has a less harmful effect on many clover species than on some erect-growing grasses and legumes.

Utilizing the leaf area concept, it might seem desirable to manage a pasture to maintain a large amount of leaf tissue so as to stay at a high forage production potential plateau. Unfortunately, pasture leaf

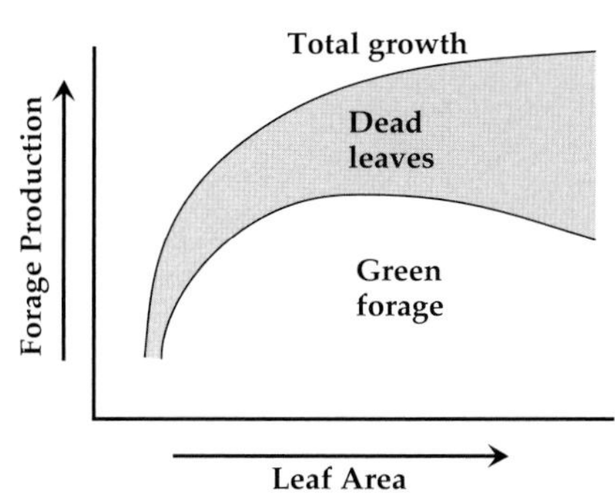

Figure 25.2. Over time, the proportion of dead leaves increases in a pasture at higher leaf area levels. The need for removal of leaves by grazing is apparent.

tissue cannot be "stored" over long periods of time because individual leaves have a limited life span of only about 30 to 60 days. Aging leaves decline in ability to utilize sunlight and finally die. Over time, the rate of leaf tissue loss to aging and decomposition increases, leaving a high proportion of dead leaves in an undergrazed pasture (**Figure 25.2**).

Grazing animals find dead leaves less acceptable and are likely to avoid them when other forage is available. Leaf growth in a pasture needs to be utilized fairly promptly, or much of it will be utilized at a less-than-optimum stage. Thus, in order to manage a pasture well it is necessary to at least periodically graze off most of the leaves to improve light penetration which stimulates growth of new leaves. Adequate light is needed to maintain an adequate supply of basal buds to develop young tillers or shoots on both grasses and legumes. Undergrazing results in heavy shading, which depresses tiller development and reduces subsequent forage production. Thus, good pasture management requires compromises with regard to plant height and closeness of grazing to achieve reasonable production of palatable and nutritious forage over a long grazing season. This can be done by manipulation of stocking density according to available forage, harvesting surplus forage for stored feed, and timely fertilization. Various forage species differ in ability to tolerate heavy or light grazing, and thus optimum pasture height varies with species.

How Animals Graze

The grazing animal is a remarkable forage harvesting machine. In one day, a mature cow may consume up to 20 percent of her weight in fresh forage. To achieve this she moves slowly over the pasture and takes successive bites (30 to 60 per minute) by drawing grass into the mouth with the tongue and then, with the grass firmly held between the tongue and the lower incisor teeth, pulling or tearing it from the plants (**Figure 25.3**).

Sheep have smaller mouthparts with which they nibble the grass, achieving a higher frequency of biting, and may bite rather than tear the selected grass by cutting it between the lower incisor teeth and the dental pad of the upper jaw. The biting action and small mouths of sheep allow more selective and closer grazing than the tearing action of cattle. Horses are highly selective grazers and prefer closely grazed forage. They often heavily graze patches in pastures, and these patches may persist from year to year unless forage availability is greatly restricted.

Grazing of cattle may occupy 6 to 11 hours per day, normally in two major periods, just before dusk and just after dawn, with shorter periods during the day or at night.

Figure 25.3. The grazing animal is a remarkable forage harvesting machine.

After a grazing period the ruminant animal rests and ruminates, regurgitating the forage, chewing, mixing it with saliva, and swallowing again. Rumination time ranges from 5 to 9 hours daily.

Effects of Grazing Animals on Pastures

The productivity and amount of each species in pastures can be rapidly and substantially altered by grazing animals. The effects may be harmful or beneficial, depending on the plants. Grazing animals defoliate, trample, and excrete dung and urine on a pasture. The interrelations of animals and plants in a pasture are complex and the effects of grazing differ greatly from hay harvesting.

1. Defoliation. Defoliation, removal of plant shoots or leaves, affects pastures differently depending on the species present, the extent of selective grazing of different plant species, the frequency of shoot removal, how much plant material is removed and how much remains, the stage of plant development, and weather conditions at time of defoliation. Defoliation is the most important influence of grazing animals on a pasture. Grazing reduces leaf area, thus affecting plant food storage, shoot development, leaf and root growth, light intensity, soil temperature, and moisture. Defoliation effects are a result of grazing pressure by livestock or by periodic mechanical clipping (**Figure 25.4**).

Heavy, close grazing when grass growing points are being elevated prior to flowering may be especially harmful to certain grasses such as orchardgrass. With tall-growing forage species, continued overgrazing generally results in weakened plants with reduced root systems, lower forage yield, greater soil erosion and water run-off, and more weeds. The closeness and frequency of defoliation often determine which forage species dominate in a pasture. For example, continuous close grazing favors bahiagrass or common bermudagrass over taller bermudagrass hybrids such as Coastal. Also, close

Figure 25.4. Overgrazing (left) adversely affects forage growth, water and soil conservation, and pasture plant persistence. Undergrazing (right) results in forage waste, reduced nutritive quality, reduced tiller development, loss of legumes, and subsequent growth.

grazing favors white clover growing in association with tall fescue. Low stocking rates or intermittent grazing allows the survival of tall species such as Johnsongrass, switchgrass, or sericea lespedeza which are not tolerant of close continuous defoliation.

Undergrazing also has undesirable effects on a pasture. Wasted forage and reduced nutritive quality are most obvious but there are other effects. Undergrazing grass-clover pastures generally decreases clover stands since the taller-growing grasses shade the lower-growing legume. For example, undergrazed tall fescue or orchardgrass will shade out ladino clover. Red clover, which is more erect, is more tolerant of grass competition.

Clovers differ in their tolerance to shade. Crimson and subterranean clovers are more shade tolerant than arrowleaf clover. Arrowleaf is particularly sensitive to shading because when shaded, buds fail to develop near the stem base for initiation of new leaves. Therefore, vigorous grazing is required to maintain good arrowleaf clover growth in late spring.

Heavy spring growth of cool season perennial grasses is frequently underutilized, resulting in shading of tiller buds. When growth slows in summer due to heat and/or drought, these pastures are often overgrazed, but recovery after rains is poor because of insufficient tillers and leaf area. Patchy grazing is a result of understocking and usually not a problem when pastures are grazed to maintain a height of 3 to 6 inches.

Some general principles on species responses to grazing are:

(1) Tall upright plants are easily defoliated by grazing and in general depend on food reserves in roots and/or stem bases for regrowth and persistence. Rotational stocking with a rest period, or continuous stocking at a rate low enough to leave a higher stubble, is required for species such as alfalfa, sericea lespedeza, pearl millet, sorghum, Johnsongrass, big bluestem, indiangrass, and switchgrass.

(2) Semi-erect species, or those having food storage areas and buds near the ground, are fairly tolerant of close grazing except under stress conditions. Species in this group include tall fescue, orchardgrass, and arrowleaf clover.

(3) Prostrate species extremely tolerant of close grazing include bermudagrass, bahiagrass, Kentucky bluegrass, white clover, and subterranean clover.

2. Treading. Treading, pugging, or pressure by hooves of grazing animals injures plants and compacts the soil, reducing pasture production (**Figure 25.5**). Pasture plants respond differently to treading damage. Ryegrass, tall fescue, Kentucky bluegrass, bermudagrass, and white clover are more tolerant of treading than small grains, orchardgrass, or red clover. Pasture treading damage is greater on wet than dry soils, on clay than sandy soils, on recently-tilled than settled soil, and with short forage as opposed to taller forage. Treading of cattle on wet clay soils substantially reduces water infiltration, resulting in less benefit from rainfall. Removing cattle from pastures during extremely wet periods can reduce treading damage.

3. Excretion. Cattle normally defecate 10 to 12 times and urinate 6 to 8 times daily. Excretion concentrates nutrients on only about 20 percent of the pasture and reduces the amount of grazable forage due to cattle avoiding dung and urine patches. Stock camps (water, shade, salt box areas) further concentrate the excreta. Separating water and salt from shade areas can reduce the problem.

Pasture Stocking Rate

The effect of a grazing system on animal productivity is strongly influenced by grazing pressure (animals per unit of forage available) and stocking rate (animals per acre). Stocking rate and its relationship to available forage are the most important management factors influencing the output of animal product from a pasture, persistence

Figure 25.5. Cattle hooves on wet pasture damage plants, compact soil, and reduce water infiltration on clay soils.

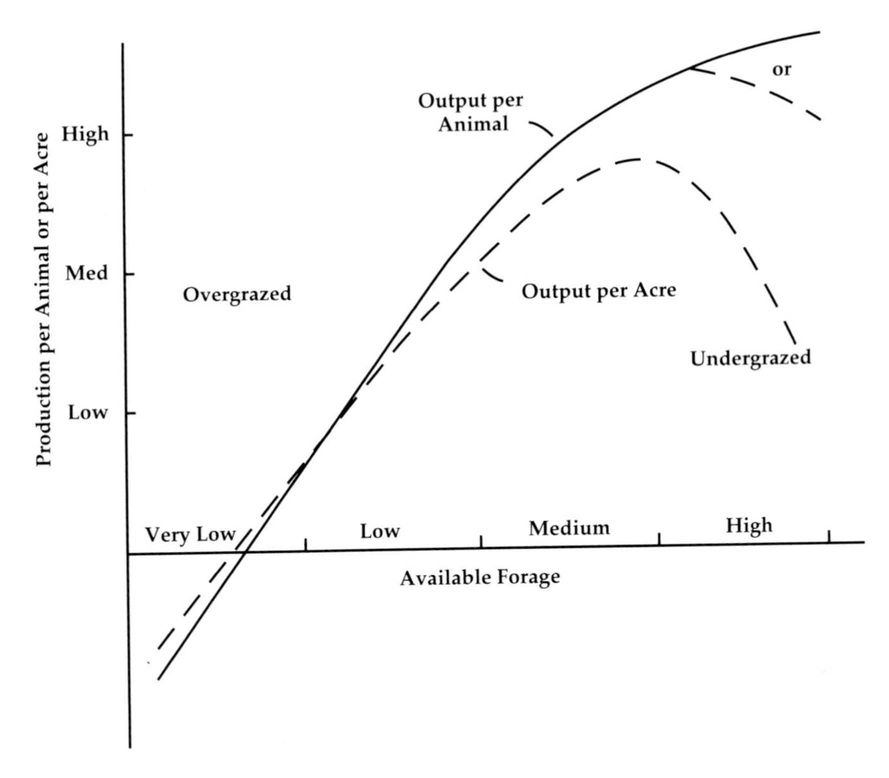

Figure 25.6. Influence of available forage on production per animal and production per acre of pasture.
Adapted from: R.E. Blaser. 1986. Virginia Agric. Exp. Stn. Bull. 86-7.

and productivity of a forage species, and financial return. Thus, stocking rates and varying the stocking density to utilize the forage on a farm influence economic returns.

Several major factors influence the optimum stocking rate.

1. Forage production. With improved forage species and good levels of fertilization, the amount of forage produced increases, creating a need to increase stock numbers in order to utilize the forage growth. At a low stocking rate, available forage and productivity per animal tend to be high, but output per acre is low (**Figure 25.6**). When pasture quality declines with accumulation of stems and dead forage, dry matter intake by animals decreases and lowers animal output.

As stocking rate is increased on a pasture, less forage is available per animal. Individual animal output decreases as animals compete for forage and have less opportunity to select. However, animal output per acre increases with stocking rate until individual animal gains are depressed to the point that the additional animals carried do not compensate for the loss. At high stocking rates, plants are weakened and forage growth is depressed.

Pastures must have sufficient leafy green forage available to allow the animals to quickly satisfy their appetites and selectively graze. This will require about 1,200 to

1,500 lb/A of dry forage on tall fescue and 1,500 to 1,800 lb/A for hybrid bermudagrass. To maintain these amounts under continuous stocking will require plant heights of about 3 to 6 inches of forage for both grass species.

In cool, moist climates such as New Zealand or in mountain areas of the southeastern USA, a good grazing recommendation for cool season grasses is to maintain a forage height of 2 to 4 inches. This encourages new leaf development and results in high quality forage. However, in most of the South where late spring, summer, and autumn drought can occur, most livestock producers prefer to maintain pastures at a somewhat greater height to have reserve forage for drought periods.

Stocking the pasture below the optimum rate for output per acre results in a higher rate of production per animal due to selective grazing. As stocking rate is increased, the opportunity for selection of high quality leafy forage is less than in pastures having a lower stocking rate. But when pastures are grazed closely, there may be more internal parasites, causing further reduced production per animal.

2. Accessibility of forage to animals. Insufficient watering points or shade may effectively reduce forage available to animals by causing overgrazing, trampling damage, and stand degradation in portions of a pasture. Separating shade, mineral, and water locations will improve utilization.

3. Nutritive value of pasture. Animals improve their diets by selectively grazing clover or leaves rather than stems of grasses. On high quality winter annual or cool season perennial grass-legume pastures with adequate available forage, selectivity is generally not a problem. Growth of new leaf tissue can be stimulated by nitrogen (N) fertilization and periodic mowing of uneaten stems.

4. Species composition of pasture. Heavy stocking rates that result in very close grazing can reduce production and the stand of some desirable forage species, allowing encroachment by weeds. Heavy stocking and close grazing of Coastal bermudagrass may result in dominance by common bermudagrass or bahiagrass. Likewise, close grazing of orchardgrass may result in invasion by less desirable species. At the southern edge of its adaptation zone, tall fescue is often invaded by bahiagrass or bermudagrass when grazed closely in summer. This problem appears to be more serious for endophyte-free than infected tall fescue. However, heavy stocking may improve summer forage quality by increasing the amount of crabgrass or white clover in endophyte-infected tall fescue pasture.

Stocking rate can be a highly useful tool for changing pasture composition. Close grazing of bahiagrass or bermudagrass in autumn makes it easier to establish annual clovers, either by sowing or natural reseeding, thus furnishing higher quality forage. Broadcast seeding of red or white clover into a dormant tall fescue pasture in late winter followed by heavy stocking of 10 to 20 animals per acre for a week or two can result in establishment of a clover stand by the "trampling" technique (see Chapter 12).

5. Seasonal variations in forage production. Pasture production varies greatly over the year and dramatically affects potential stocking rate (**Figure 25.7**). Stocking rate adjustment is essential if forage is to be efficiently utilized and animal production maintained over a long period. Surplus forage can be harvested from ungrazed pastures during peak seasons of growth. Regardless of grazing management, stored feed will likely be needed during periods of limited growth. The requirement for stored feed can be reduced by having both a cow-calf and a stocker operation. This situation gives more flexibility because, if necessary, animals can be bought or sold to increase or reduce stocking rate. However, prices are often depressed during periods of low forage growth. Astute timing in sale of cattle is critically important.

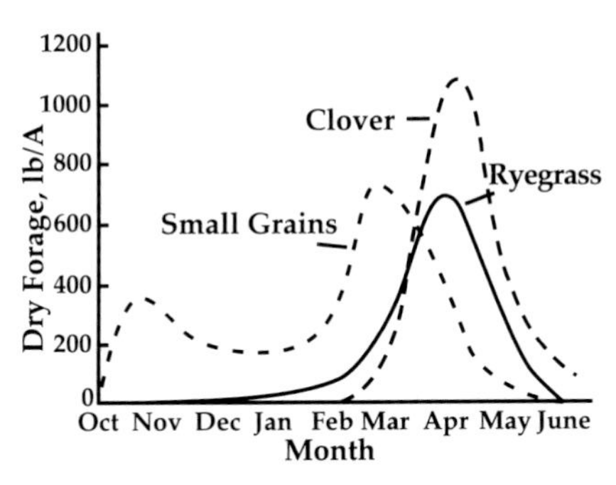

Figure 25.7. Large seasonal differences in forage production require adjustments in stocking rate to utilize forage as illustrated by winter annual pasture growth in South Alabama.

6. Economic considerations. Stocking rates which maximize the value per acre from meat-producing animals will be lower than stocking rates which maximize animal gain per acre. Fixed and variable costs per acre rise sharply with increasing stocking rates. Thus, the margin over variable and fixed costs is likely to be greater for the livestock system when stocking rates are considerably lower than those required to maximize animal output per acre.

Grazing Methods

The objective of a grazing method should be to manage the pasture and other feed inputs to efficiently produce animal products. **Remember, managing forage quantity and quality over the grazing season is of much greater importance than which grazing method is employed.** A number of different grazing methods are successfully used for efficient livestock production in the South.

Several methods of grazing have been developed to accomplish various objectives. The advent of lower cost, higher voltage electric fences has made it more feasible to implement various grazing methods which reduce pasture waste, conserve surplus forage as hay or silage, and possibly increase forage quality for the grazing animals. It should be remembered that the method of grazing used generally has less influence on production than the stocking rate.

Grazing methods are of two general types–continuous and several variations of intermittent grazing. They vary from simple to complex (**Figure 25.8**).

Continuous Stocking (often referred to as Continuous Grazing). Animals are maintained on a single pasture unit during the grazing season. This method allows animals to selectively graze, unless the stocking rate is too high. Unless animal numbers or pasture size is adjusted as pasture conditions change, this method may result in some plants being undergrazed and others being overgrazed. Also, without adjustment, the stocking

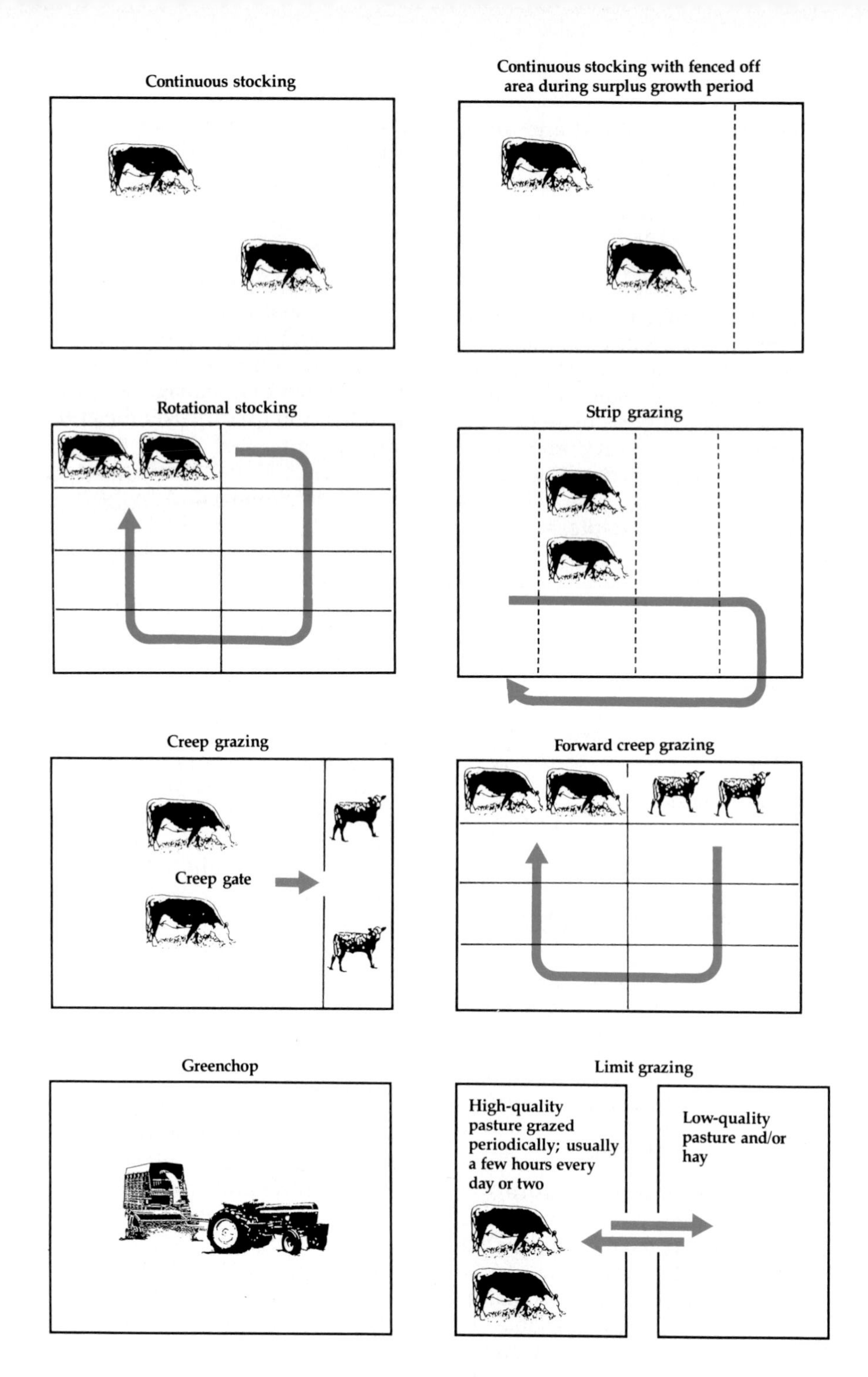

Figure 25.8. A variety of grazing/harvest methods may be used successfully.

rate may be near ideal part of the time and too high or low at other times during the grazing season. Animals may be **set-stocked** with a fixed number of animals, or numbers may be adjusted to fit the feed supply over the grazing season. Adjustment of the stocking rate as needed greatly improves forage utilization. A temporary fence is sometimes used to close off part of a continuously stocked pasture during periods of surplus growth. This allows accumulated forage to be harvested for stored feed.

Continuous stocking has often been incorrectly equated with overgrazing and poor management. A continuously stocked pasture can be just as productive and efficient as any other method provided that available forage is controlled by adjusting stock numbers as needed. Continuous stocking encourages a high number of grass tillers to improve pasture stability and production over a long period of time.

Certain plant species such as alfalfa, sericea lespedeza, Johnsongrass, big bluestem, indiangrass, and switchgrass are not suited to continuous stocking unless the forage is allowed to remain tall. Stands of conventional hay-type alfalfa varieties will weaken and die under continuous close stocking. However, new grazing-tolerant alfalfa varieties will persist under moderate continuous stocking.

Rotational Stocking (often referred to as Rotational Grazing). The pasture is subdivided into several paddocks and for a particular subdivided area, a rest period follows each grazing period **(Figure 25.9).** The number of paddocks may vary from only a few to 12 or more. A high stocking rate is imposed on a paddock for a short time, often 1 or 2 days but generally no more than 7 to 10 days, and the animals are then shifted to another paddock. During periods of surplus growth, one or more paddocks may be harvested for hay while animals are rotated through other paddocks more quickly.

Results of many pasture experiments with steers or heifers in the Southern USA have failed to show a consistent advantage for rotational stocking **(Table 25.1).** The effect on gain per acre ranged from minus 22 percent to plus 41 percent. In most of these experiments conducted over a number of years, stocking rates in all methods were adjusted by adding or subtracting steers to maintain similar forage availability. Average daily gain of animals was either unaffected or decreased by rotational stocking.

Rotational stocking, especially at higher stocking rates, reduces the opportunity for animals to select only the leafiest forage, thus decreasing the quality of forage ingested, and sometimes lowering individual animal gains. This is especially true on low-quality warm season perennial grasses. Small grains and ryegrass are highly palatable and the nutritive quality remains high over the grazing season. Thus, animals on these high-quality pastures can maintain high performance regardless of grazing method.

A variation on rotational stocking is to give first access in a paddock to animals having high nutritional requirements such as growing steers and lactating dairy cows. This is referred to as **"forward grazing."** For example, growing animals with high potential for gain can be allowed to graze a paddock first, followed by beef cows which utilize the remaining lower-quality

Figure 25.9. Rotational stocking of endophyte-free tall fescue can give better control over the pasture and should reduce forage waste.

forage. This method maintains high gains on the steers and removes the old residue to stimulate new growth of high-quality leafy forage while the pasture is rested. This approach can also be used with lactating dairy cows followed by dry cows or replacement heifers.

Some Benefits of Rotational Stocking. In addition to potential increased carrying capacity, there are a number of other benefits of rotational stocking.

(1) Better control over livestock. Animals are often more docile, easier to handle, and can be observed and examined frequently by the owner.

(2) Better persistence and productivity of pasture plants that are sensitive to close continuous grazing. Plants such as alfalfa, sericea lespedeza, and endophyte-free tall fescue will usually benefit from rotational stocking. When a mixture of endophyte-free tall fescue and common bermudagrass was grazed year around with beef cows and calves in Georgia, rotational stocking maintained better stands and productivity of the tall fescue, improving carrying capacity and calf gain per acre, and decreasing the amount of hay fed in winter by 29 percent **(Table 25.2)**.

(3) Improved utilization of more species in the pasture and less waste. There are indications that more weeds are eaten, favoring dominance by desirable species.

(4) Concentrating animals in smaller areas for short periods allows successful "trampling in" of broadcast-seeded clover.

(5) Moving livestock from one paddock to another generally causes the producer to more carefully evaluate pasture species and productivity, thus becoming a better manager.

Layout and utilization of grazing paddocks

(1) Paddock size, location, and arrangement. Rotational stocking will be more costly because of additional fencing, stock water locations, and access roads. Paddock size will depend on the forage available per acre that will be needed to

Table 25.1. **Beef steer performance as influenced by continuous and rotational stocking or strip grazing of pasture and greenchop feeding in the Southern USA.**

Pasture species	Location	Method	Daily gain per steer, lb	Change from continuous stocking, %	Total gain per acre, lb	Change from continuous stocking, %
WARM SEASON SPECIES						
Bermudagrass + N	South Carolina Piedmont[1] 3 yr	Continuous	1.37		738	
		Rotational (4 paddocks)	1.27	−7	749	+1
Bermudagrass + N	South Georgia[2] 3 yr	Continuous	1.31		535	
		Rotational (4 paddocks)	0.99	−24	418	−22
		Strip	0.86	−34	434	−19
		Greenchop	0.81	−38	577	+8
Bermudagrass + N	Western Arkansas[3] 2 yr	Continuous	1.54		565	
		Rotational (10 paddocks)	1.31	−15	686	+21
Bermudagrass + N	North Florida[4] 2 yr	Continuous	1.07		638	
		Rotational (15 paddocks)	1.01	−6	649	+2
Low-tannin sericea lespedeza	North Alabama[5] 3 yr	Continuous	1.87		306	
		Rotational (3 paddocks)	1.65	−12	276	−10
COOL SEASON SPECIES						
Orchardgrass + N	Virginia[6] 3 yr	Continuous	1.30		364	
		Rotational (4 paddocks)	1.23	−5	388	+8
Endophyte-free tall fescue, orchardgrass, Kentucky bluegrass	Central Missouri[7] 2 yr	Continuous	1.51		211	
		Rotational (10 paddocks)	1.42	−6	298	+41
Endophyte-free tall fescue + N	Western Arkansas[3] 3 yr	Continuous	1.63		356	
		Rotational (10 paddocks)	1.34	−18	429	+20
Endophyte-infected tall fescue + N	Western Arkansas[3] 3 yr	Continuous	1.17		308	
		Rotational (10 paddocks)	0.98	−16	384	+25
Endophyte-free tall fescue + N	Central Georgia[8] 3 yr	Continuous				
		1.3 steers/A	2.20		264	
		2.0 steers/A	2.12		378	
		Rotational (8 paddocks)				
		1.5 steers/A	2.45	+11	333	+26
		1.9 steers/A	1.95	−10	337	−11
Ryegrass, crimson clover + N	West Louisiana[9] 2 yr	Continuous	2.37		675	
		Rotational (15 paddocks)	2.29	−3	828	+23
Wheat, ryegrass + N	Central Mississippi[10] 3 yr	Continuous	2.15		746	
		Rotational (6 paddocks)	1.72	−20	733	−2
Rye, wheat, ryegrass + N	Central Alabama[11] 1 yr	Continuous	1.69		294	
		Rotational (10 paddocks)	1.66	−2	302	+2

[1]R.F. Suman, S.G. Woods, T.C. Peele, and E.G. Godbey. 1962. Agron. J. 54:26-28.
[2]R.H. Hart, W.H. Marchant, J.L. Butler, R.E. Hellwig, W.C. McCormick, B.L. Southwell, and G.W. Burton. 1976. J. Range Manage. 29:372-375.
[3]L.M. Tharel and M.A. Brown. 1995. Unpublished data.
[4]B.W. Mathews, L.E. Sollenberger, and C.R. Staples. J. Dairy Sci. 77:244-252.
[5]S.P. Schmidt, C.S. Hoveland, E.D. Donnelly, J.A. McGuire, and R.A. Moore. 1987. Alabama Agric. Exp. Stn. Cir. 288.
[6]R.E. Blaser. 1986. Virginia Agric. Exp. Stn. Bull. 86-7.
[7]J. Gerrish, S. Marley, and R. Plain. 1992. Proc. Amer. Forage Grassl. Conf. p. 147-151.
[8]C.S. Hoveland, M.A. McCann, and N.S. Hill. University of Georgia. Unpublished data.
[9]G.D. Mooso, D.G. Morrison, C.C. Willis, and J.E. Miller. 1988. Louisiana Agric. 32. No. 3. p. 8-10.
[10]E.G. Morrison. 1990. Mississippi Agric. Exp. Stn. Inf. Bull. 167.
[11]D.I. Bransby, D.D. Kee, and W.H. Gregory. 1989. Alabama Agric. Exp. Stn. Highlights of Agric. Res. 36. No. 4.

meet the feed requirements of the animals grazing during the days on that paddock (**Appendix A.28**). Uniform sized paddocks with parallel sides of similar length are desirable (see Chapter 24). Paddock boundaries should follow land contours and changes in soil type and slope to allow growth of forage species best suited to the site. Access to water is best achieved by an alleyway which connects to each paddock. A wagonwheel design with the paddocks narrowing to one central water source should not be used in high rainfall areas such as the South as muddy conditions near the water source will result.

(2) Paddock number. It would appear to be ideal to move animals daily to a new paddock to obtain high quality forage, less waste, uniform regrowth, and good manure distribution. Realistically, grazing

Table 25.2. **Effect of year-around continuous vs rotational (12 paddocks with cattle rotated every 2 days) stocking of endophyte-free tall fescue and common bermudagrass mixed grass pastures in the Piedmont of Georgia, 3-year average.**

	Continuous	Rotational	Change, %
Stocking rate, cow-calf units/acre	0.50	0.69	+38
Calf weaning weight, lb	490	486	0
Total calf gain/acre, lb	243	334	+37
Cow pregnancy rate, %	93	95	0
Hay fed/cow, lb	2,430	1,680	−31

Source: C.S. Hoveland, M.A. McCann, and N.S. Hill, University of Georgia. Unpublished data.

periods should be less than four days to obtain these benefits. The rest interval needed between grazings depends on the pasture species and growing conditions. Most grass species are fairly tolerant of the rest interval while alfalfa will need 3 to 5 weeks to replenish root carbohydrates. Generally, 6 to 8 paddocks can accomplish this but a few more can give greater flexibility and control. Since single-strand electric fence can be used, fencing can be moved as needed to adjust the system.

(3) Movement of animals between paddocks. Effective rotational stocking should be **flexible** so that rotation among paddocks is not based on a fixed time schedule but rather on available forage and rate of forage growth. During periods of surplus growth, one or more paddocks may be harvested for hay or haylage while animals are rotated through other paddocks more quickly. With experience, a manager can use forage height to estimate the amount of forage present (**Appendices A.9 and A.28**) and subsequently calculate how long it will feed the animals in the paddock. Forcing high-producing animals to overgraze a paddock will result in poor animal performance.

Creep Grazing. This method allows young calves or lambs to pass through a fence opening (creep) to a special small pasture of higher quality forage (such as pearl millet, small grains, clover, or alfalfa) than the lower quality pasture where their mothers are maintained. This is particularly effective in summer when low forage availability restricts calf gains and when nutritive quality of the perennial grass pasture is low (**Table 25.3**). Calf gain increases and often cow condition is improved. This is a relatively inexpensive system as it

Table 25.3. **Creep grazing of beef calves on pearl millet from June-September (104 days) when cow-calf pairs were maintained on tall fescue pasture in northern Alabama.**

	Control	Creep-grazed
Calf weight gain, lb	144	219
Calf average daily gain, lb	1.38	2.10
Cow weight change, lb	−60	+27

Source: E.E. Thomas, J.T. Eason, D.M. Ball, and B.G. Ruffin. 1983. Alabama Agric. Exp. Stn. Highlights Agric. Res. Vol. 30, No. 2.

Figure 25.10. A creep gate allows calves, but not their mothers, to have access to higher quality pasture.

requires only a small area of high quality pasture for the young growing animals. Creep pastures should be close to water or loafing areas so calves will utilize the high quality forage.

Forward Creep Grazing. This method allows calves or lambs with a high nutrient requirement to pass through an opening in the fence to graze fresh ungrazed pasture ahead of their mothers in a conventional rotational method. A creep gate is illustrated in **Figure 25.10.**

Strip Grazing. Movable electric fence is used ahead of and behind the animals to ration daily forage. This method results in high forage utilization and is most effective with excellent pasture during cool weather when nutritive quality declines more slowly. Rapid grazing over the pasture and harvesting of surplus growth are essential or else forage quality will deteriorate. With low-quality forage, animal daily gains may suffer with strip grazing because of reduced selectivity (**Table 25.1**). This method is not often economical because of labor, water, shade requirements, and fencing costs.

Deferred Grazing or Stockpiling. This method consists of delaying grazing during part of the grazing season, usually autumn, to provide forage during winter or a dry season when growth is greatly reduced. Quality of forage deteriorates, but this technique can be useful in reducing the amount of hay feeding. It works especially well with tall fescue in the upper South. Forage quality of most warm season grass species deteriorates quickly after frost, reducing the value of such a program.

Limit Grazing. Maintaining livestock on lower quality pasture, but allowing them access to a high quality winter annual pasture for a few hours every few days, can reduce waste from trampling. This is most often done with animals moved from frosted warm season perennial pasture to winter annual pasture.

Greenchop. Forage is chopped mechanically and fed green to livestock. This method reduces waste by grazing animals so that more animals can be fed per acre, but forage selectivity is reduced and individual animal output is often lower (**Table 25.1**). Equipment costs are high with this method. Forage often is not harvested at optimum growth stage, lowering nutritive quality. Also, daily harvesting can be difficult when rain results in wet soils and poor field conditions.

Summary

Much skill is required to obtain optimum or near optimum animal production from pastures. It is one thing to produce forage, but efficient utilization is usually a greater challenge. In order to provide adequate quantities of good quality forage and economically convert it into animal products, good grazing management is essential. Good grazing management involves frequent observation of pasture height, and periodic adjustment of stocking rate or movement of animals as needed.

The choice of grazing method to be used depends on the individual farm and the livestock producer. Continuous and some form of intermittent stocking such as rotational or creep grazing may be used on the same farm for different livestock enterprises or at different seasons of the year. Different methods are not mutually exclusive, and one is not necessarily superior to another.

A grazing method is a tool that allows a producer to efficiently and profitably harvest the forage with livestock and maintain the pasture in a productive state. Each method requires management control to be successful. This involves variable stocking rates which may be achieved by altering animal numbers per acre, or altering the size of the land area to a fixed number of animals, harvesting surplus forage as hay and/or mowing excess growth in a pasture.

Every grazing method has advantages and disadvantages. A producer must select the one which best suits a particular situation. This varies, depending on the type and class of livestock, pasture species, resources available, and objectives of the operation. ■

Chapter 26

Forages for Beef Cows and Calves

BEEF CATTLE production in the South consists primarily of commercial and purebred cow-calf operations. Commercial cow-calf producers normally sell calves at weaning which are subsequently transported to other areas of the USA where they are pastured and then finished in a feedlot. However, an increasing number of Southern producers are purchasing weanling calves or retaining calves from their cow-calf operations for stockering on high quality pasture to benefit from the attractive profit potential from sale of feeder calves.

There are over 12 million beef cows in the South, including eastern Texas and Oklahoma. Most of the herds are small, often part-time operations or associated with crop farming. A common problem is uncontrolled breeding, resulting in calves being born throughout the year. Due to this and other poor management practices, calf death rates are increased and calf weaning weights are low, often much less than 500 lb. Poor quality forage during much of the year also contributes to low calf weaning weights and reduced cow reproduction. See **Table 26.1**.

Nutrient Needs

The primary objectives in efficient cow-calf production should be to: (1) maintain the cow as cheaply as possible; (2) obtain good reproduction; and (3) attain high calf

Figure 26.1. Crossbred cows and calves grazing Coastal bermudagrass near Malakoff in east Texas.

Table 26.1. Estimated impact on average calf weaning weights if Southern beef producers used near-optimum management.[1]

Item	Increase, lb
Nutrition	50-75
Genetics	30-35
Implants	25-30
External parasite control	15-20
Internal parasite control	15-20

[1]Based on estimates by university agronomists and animal scientists.
Source: D.M. Ball

weaning weights. To achieve these objectives, it is necessary to produce forage in pastures and hay fields at low cost and of sufficient quality to meet the needs of the animals with little supplemental feeding.

A controlled breeding season is necessary to insure adequate nutrition for the cow during early lactation and to match available supplies of high quality pasture to the growing calf. Dry, pregnant mature cows can be maintained on relatively low quality forage having 7 to 8 percent crude protein and 50 percent digestible dry matter. Lactating cows require a diet furnishing about 10 to 12 percent crude protein and 60 percent digestibility.

An important factor affecting cow herd production, rebreeding, and calving percentage is the available energy level during the first three months of lactation (see Chapter 17). Even more important is the condition of the cow or heifer at the time of calving. Cows calving in good body condition usually have no trouble rebreeding and raising a good calf when provided with adequate nutrition. However, cows in thin body condition at calving will have extremely low rebreeding rates even when fed very high levels of nutrients.

Ideally, calving should occur 4 to 6 weeks prior to the initial availability of good quality pasture to correspond with peak beef cow nutrient demands and allow nursing calves to use pasture for much of their growth. Thus, the ideal calving period differs in various zones of the South depending on pasture availability.

Autumn calving is desirable in Zone A (See **Figure 3.2**) where limited small grain or other winter annual pasture is available to cows, or creep grazing is available to calves. If only warm season perennial grasses are used, late winter calving is more desirable. In Zone B, where tall fescue-legume pasture is available, winter calving is more suitable, while in Zones C and D late winter or even spring calving may best match supplies of pasture to animal needs. In any of these areas, summer is generally the single worst time for a cow to calve.

Overfeeding dry cows, with resulting weight gain, can result in decreased fertility and increased calving problems. The period from weaning until calving is the portion of a cow's reproductive cycle during which it is easy to improve her body condition. Her nutrient requirements are low and she can fatten on late summer-early autumn low-quality grass.

Although it is a common practice to provide protein supplement with hay feeding, such supplementation is usually not needed if the hay was harvested at early maturity or if it contains clover, alfalfa, or birdsfoot trefoil. Salt should be provided and supplemental phosphorus (P) may be needed. Most cool season grasses contain adequate P levels, but warm season grasses may be deficient. When supplying P, some calcium (Ca) should be included in the mineral supplement. Some areas in the South also are deficient in selenium and thus require supplementation.

Possible Forage Systems

Various pasture-hay systems have been effectively used for feeding beef cow herds (**Table 26.2**). All of these forage systems except endophyte-infected tall fescue plus

Table 26.2. Performance of beef cows and calves on year-around pasture-hay systems in the South.

Pasture	Winter hay	Location	Nitrogen fertilizer, lb/A	Land/ cow-calf, acres	Calf daily gain, lb	Calf weaning weight, lb	Calf gain/ acre, lb	Cow pregnancy rate, %
WARM SEASON GRASSES								
Coastal bermuda-grass + arrowleaf clover	Coastal bermuda	East Texas[1]	0	1.10	2.70	645	446	95
Coastal bermuda-grass + arrowleaf clover	Coastal bermuda + 134 lb/cow of liquid protein	Alabama Piedmont[2]	100	2.00	1.72	594	209	97
Coastal bermuda-grass	Coastal bermuda + 294 lb/cow of cotton-seed meal	South Alabama[3]	200	0.94	1.53	453	458	95
Dallisgrass + white clover + caley peas	Dallisgrass	Alabama Black Belt[4]	0	2.00	1.70	484	242	—
COOL SEASON GRASSES								
Endophyte-free tall fescue + ladino clover	Sericea lespedeza	Central Georgia[5]	0	1.75	1.81	499	315	94
Endophyte-free tall fescue + ladino clover	Tall fescue-red clover hay + stockpiled tall fescue-red clover pasture	Northern Virgnia[6]	0	1.78	1.83	551	310	94
Endophyte-infected tall fescue	Orchardgrass hay + corn silage	Southern Indiana[7]	100	1.80	1.28	353	—	71
Endophyte-infected tall fescue + ladino and red clover	Orchardgrass hay + corn silage	Southern Indiana[7]	0	1.80	1.83	436	—	92
Orchardgrass	Orchardgrass hay + corn silage	Southern Indiana[7]	100	1.80	1.79	430	—	90

1F.M. Rouquette, Jr., M.J. Florence, V.A. Haby, and G.R. Smith. 1990. Texas A&M Univ. Overton Res. Ctr. Tech. Rep. 90-1. p. 153-165.
2C.S. Hoveland, W.B. Anthony, J.A. McGuire, W.A. Griffey, and H.E. Burgess. 1977. Alabama Agric. Exp. Stn. Bull. 497.
3R.R. Harris, V.L. Brown, W.B. Anthony, and C.C. King, Jr. 1970. Alabama Agric. Exp. Stn. Bull. 411.
4C.C. King, Jr., W.B. Anthony, S.C. Bell, L.A. Smith, and H. Grimes. 1971. Alabama Agric. Exp. Stn. Bull. 424.
5J.A. Stuedemann and C.S. Hoveland, USDA and Univ. of Georgia. Unpublished data.
6V.G. Allen, J.P. Fontenot, D.R. Notter, and R.C. Hammes, Jr. 1992. J. Anim. Sci. 70: 576-587.
7D.C. Petritz, V.L. Lechtenberg, and W.H. Smith. 1980. Agron. J. 72:581-584.

N resulted in satisfactory cow pregnancy levels, calf weaning weights, and calf daily gains. The poor performance on endophyte-infected tall fescue in southern Indiana occurred even though winter hay feeding was supplemented with corn silage for about 2 months after calving and prior to grazing in mid-May. Growing clover with endophyte-infected tall fescue resulted in animal performance similar to that on orchardgrass pasture. Endophyte-free tall fescue plus clover pastures with winter feeding of tall fescue plus clover hay in Virginia, or endophyte-free tall fescue/clover pasture plus sericea lespedeza hay in Georgia, both gave excellent calf weaning weights.

Legumes in grass pastures gave high calf weaning weights although generally requiring more acreage per cow-calf unit than grass-N systems. Maintaining cows on bermudagrass pasture and hay in Alabama gave good performance at less than 1 acre per cow. However, it was a costly program since it required high N fertilization and winter feeding with cottonseed meal supplement because the hay contained only 9 percent crude protein. A cheaper system of Coastal bermudagrass with arrowleaf clover and no N fertilizer or supplement resulted in high calf weaning weights in east Texas.

Year-round pasture-hay systems differ considerably, depending on the climatic zone (**Figure 3.2**). A number of possible forage systems are outlined for various zones:

Zone A.

(1) Bermudagrass, bahiagrass, or dallisgrass pastures overseeded with an annual clover, rye and/or ryegrass. Surplus hay harvested from fenced-off areas during late spring and summer. About 1.5 to 2 acres needed per cow if no crop residues are available.

(2) Hay or crop residues such as cornstalks, peanut hay, or soybean residue fed in midwinter.

(3) Autumn-born calves creep grazed in winter on rye, wheat, or oats, and/or ryegrass planted early on prepared land.

Zone B.

(1) Tall fescue, preferably planted with white or red clover, grazed from early autumn until early summer. A tall fescue/common bermudagrass mixture with N fertilization in autumn and late winter can provide grazing over most of the year. About 1.5 acres needed per cow. Sericea lespedeza, tall fescue, or bermudagrass hay fed in midwinter. Rye overseeded on bermudagrass or sericea for grazing. Surplus tall fescue hay can be harvested from fenced areas in spring.

(2) Sericea or bermudagrass hay harvested in late spring. About 0.5 acre needed per cow. Sericea or bermudagrass grazed during mid to late summer when tall fescue is unproductive.

Zone C.

(1) Tall fescue, orchardgrass, or Kentucky bluegrass with red or white clover or birdsfoot trefoil, grazed from late winter until late autumn. About 1.5 to 2 acres needed per cow. Surplus growth harvested for hay in spring.

(2) Stockpiled tall fescue grazed in early winter. Hay fed in midwinter. About 0.5 acre of sericea per cow could be substituted for part of the grass acreage to provide hay and midsummer grazing during drought.

Zone D.

(1) Tall fescue, orchardgrass, Kentucky bluegrass, or timothy, each planted with red or white clover or birdsfoot trefoil, grazed from early

Figure 26.2. Endophyte-free tall fescue and white clover make an outstanding pasture for beef cows and calves.

spring until autumn. About 1.5 to 2 acres needed per cow. Surplus growth harvested for hay in spring.

(2) Stockpiled tall fescue grazed in late autumn. Hay fed in winter.

On many farms in all zones, crop residues can be utilized as additional forage for cattle. These materials may reduce the acreage needed for hay production. Normally, winter is the main hay feeding period, but severe summer droughts may require additional feeding. Generally, about 1 to 1.5 tons of hay per cow annually should be adequate for winter feeding.

Sorghum-sudan or pearl millet is sometimes planted as supplemental summer grazing or hay for beef cows and calves. These grasses are much more expensive than perennial pasture and cannot be justified except in emergency situations or as creep grazing for calves.

Summary

Forage systems differ from one zone to another and within zones depending on land capabilities and type of enterprise. Regardless of the forage system, it is important to determine seasonal distribution of pastures on a farm and adjust calving to match cow and calf needs to forage quality. During periods of surplus growth, cattle should be restricted to a smaller area to allow harvesting of hay at the proper stage of maturity. Pastures should be used as much of the year as possible to reduce the cost of supplemental feeding. ■

Chapter 27

Forages for Beef Stockering

STOCKER ANIMALS are grazed on high quality pasture from weaning to usually 700 or 800 lb before being sold to feedlots for finishing. However, only an estimated 15 to 20 percent of the calves are retained in the South and grown to these heavier weights. The greatest opportunity for improving profitability in Southern beef production lies in stockering (also referred to as backgrounding) weaned calves on high quality pasture.

Historically, the price of calves is lowest in early autumn when the greatest number of calves are available. This is the time of year when most cow-calf producers wean their calves. Calf prices vary with weight. Usually, lightweight calves are priced higher than heavy calves. Prices for calves typically increase in winter and peak in March and April. The price of calves sold in late spring may be slightly lower due to the weight gains on winter pasture. However, the cost of this late spring gain is relatively low, generally making it a profitable enterprise in most years.

The large supply of available calves in the region offers an opportunity for persons wishing to concentrate on this intensive system of livestock production. However, stockering is a high-risk and capital-intensive enterprise compared with a cow-calf operation. Purchasing and selling of cattle, health care, production and grazing management of high quality pastures, and supplemental feeding require a high degree of skill and experience. Generally, only larger operations (usually at least one truckload, about 60 calves) can justify the managerial input required. Economy of size is also needed to spread fixed investments over a large number of animals, acreage of land,

Figure 27.1. Beef steers grazing rye, ryegrass, arrowleaf clover pasture in the Black Belt of central Alabama.

Table 27.1. Beef steer performance on pastures in the Southern USA.

Pasture species	Location	Nitrogen per acre, lb	Average daily gain, lb	Gain per acre, lb	Gain per animal, lb	Grazing season, days
Cool season annuals						
Rye + ryegrass + arrowleaf clover	Alabama Piedmont[1]	100	1.98	442	394	193
Ryegrass + arrowleaf clover	Western Louisiana[2]	178	2.21	520	370	149
Wheat + ryegrass	Central Missisippi[3]	122	2.19	609	404	180
Cool season perennials						
Tall fescue (endophyte-free)	Alabama Black Belt[4]	200	1.82	426	322	171
Tall fescue (endophyte-infected)	Alabama Black Belt[4]	200	1.00	301	174	171
Tall fescue (endophyte-free) + birdsfoot trefoil	North Alabama[5]	0	2.37	550	384	142
Orchardgrass + ladino clover	North Alabama[6]	0	1.60	576	374	183
Alfalfa	North Alabama[7]	0	2.16	475	349	163
Warm season annuals						
Pearl millet	South Georgia[8]	150	1.26	366	—	86
Sorghum-sudan	North Alabama[9]	100	1.10	210	84	77
Warm season perennials						
Low-tannin sericea	North Alabama[7]	0	1.65	276	245	139
High-tannin sericea	North Alabama[7]	0	1.39	248	190	139
Coastcross-1 bermudagrass	South Georgia[10]	150	1.50	469	—	131
Coastal bermudagrass	South Georgia[10]	150	1.08	331	—	131
Pensacola bahiagrass	South Georgia[10]	150	0.95	222	—	131

[1]W.B. Anthony, C.S. Hoveland, E.L. Mayton, and H.E. Burgess. 1971. Alabama Agric. Exp. Stn. Cir. 182.
[2]C.P. Bagley, J.I. Feazel, and K.L. Koonce. 1988. J. Prod. Agric. 1:149-152.
[3]E.G. Morrison. 1990. Mississippi Agric. Exp. Stn. Inf. Bull. 167.
[4]C.S. Hoveland, S.P. Schmidt, C.C. King, Jr., J.W. Odom, E.M Clark, J.A. McGuire, L.A. Smith, H.W. Grimes, and J.L. Holliman. 1983. Agron.J. 75:821-824.
[5]C.S. Hoveland, R.R. Harris, R.L. Haaland, J.A. McGuire, W.B. Webster, and V.H. Calvert II. 1985. Alabama Agric. Exp. Stn. Bull. 567.
[6]R.R. Harris, C.S. Hoveland, J.A. McGuire, W.B. Webster, and V.H. Calvert II. 1984. Alabama Agric. Exp. Stn. Bull. 563.
[7]S.P. Schmidt, C.S. Hoveland, E.D. Donnelly, J.A. McGuire, and R.A. Moore. 1987. Alabama Agric. Exp. Stn. Cir. 288.
[8]G.M. Hill, W.W. Hanna, and H.D. Wells. 1989. Proc. 16th Int. Grassl. Cong. Nice, France. pp. 1173-1174.
[9]C.S. Hoveland, R.R. Harris, J.K. Boseck, and W.B. Webster. 1971. Alabama Agric. Exp. Stn. Cir. 188.
[10]P.R. Utley, H.D. Chapman, W.G. Monson, W.H. Marchant, and W.C. McCormick. 1974. J. Anim. Sci. 38:490-495.

and equipment. The investment is large but profit potential is good for well-managed enterprises.

Nutrient Needs

Profitable stockering of young, growing cattle requires an average daily gain of at least 1.5 lb for the season. High forage quality is essential for high daily gains and profitability of steers. A growing beef steer or heifer requires forage with about 12 percent crude protein and 65 to 68 percent digestibility in order to make such gains (see Chapter 17). If possible, an ionophore should be fed to steers on pasture. These compounds increase daily gains approximately 10 percent and reduce the potential for bloat.

Steer performance varies greatly on different pasture species in the South (**Table 27.1**). In addition to individual daily gain, the carrying capacity and length of grazing season of a pasture system are extremely important.

Warm season perennial grasses such as bermudagrass and bahiagrass have generally furnished poor average daily gains because of low and slowly digestible energy. Although excellent daily gains can be obtained on bermudagrass in spring, animal performance declines rapidly during summer (**Figure 27.2**). More digestible varieties such as Grazer are somewhat better in late summer. Nitrogen (N) fertilization and mowing to remove stems and old leaves have been helpful in improving forage quality and animal performance of bermudagrass.

Warm season annual grasses such as pearl millet and sorghum-sudangrass are highly productive and have allowed high stocking rates. But, because of relatively low daily gains, the costs are high. Stockering

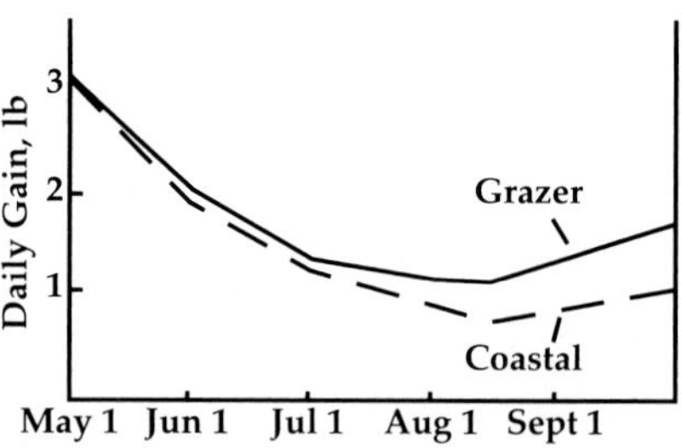

Figure 27.2. Beef steer average daily gain was high on bermudagrass in spring but declined sharply during summer in north Louisiana.
Adapted from: B.B. Greene, M.M. Eichorn, W.M. Oliver, B.D. Nelson, and W.A. Young. 1990. J. Prod. Agric. 3:253-255.

of calves on warm season annuals has generally been about a break-even proposition at best. Calves for such a system are purchased in early spring when prices are normally high and sold in late summer when prices are at their lowest. Unless this fits into a program of retained ownership through a feedlot finishing period, making a profit is difficult.

Small grain (rye, wheat, oats) and/or annual ryegrass pastures have traditionally been used in the lower South for most beef stocker production. Mixtures of rye, ryegrass, and arrowleaf clover can provide high gains over a longer grazing season. However, winter annual pastures are fairly expensive to grow and must be well managed to produce profits. Cost of gains have generally been in the range of $0.25 to $0.40 per pound during normal winters.

Orchardgrass alone, or with a clover, has furnished high daily gains in areas where it is adapted. A problem with this cool season perennial grass is the relatively short stand life under close grazing, particularly in the mid-South area.

Figure 27.3. Good clover can improve the quality of perennial grass pastures for stocker animals.

Tall fescue generally gives low daily steer gains because most pastures are heavily infected with a fungal endophyte (see Chapter 23). Endophyte-free tall fescue has given excellent steer gains, approaching that of winter annuals. The high daily gains, high stocking rate, and long grazing season of this grass make profitable stockering possible over much of the South in areas where stands can be maintained. With a perennial sod, trampling and mud problems during wet winters will not be as great compared to annual species planted in a prepared seedbed.

Including a legume such as alfalfa, red clover, or ladino clover with endophyte-free tall fescue may further increase gains (**Figure 27.3**). Costs are also reduced due to N supplied by the legumes. If legumes make up a good proportion of the forage, endophyte-infected tall fescue pastures can produce good stocker gains.

Alfalfa offers potential for stockering of steers during spring, summer, and autumn. Daily gains of 2 lb per day can be obtained on this pasture during spring and summer. Rotational stocking, to allow rest periods for accumulation of root food reserves, is essential to maintain stands and productivity on hay-type alfalfa varieties (see Chapter 25).

Possible Forage Systems

Zone A.

(1) Rye, wheat, or oats planted on prepared land in September provides grazing from mid-October or November until mid-April. Adding annual ryegrass and arrowleaf clover can extend the grazing season to June. Ryegrass alone can be used in the Gulf Coast area. Supplemental hay, silage, or other feedstuffs of similar quality should be fed in January when pasture growth is restricted. Stocking rate of 1.5 to 2 steers per acre.

(2) Pearl millet or a sorghum-sudan hybrid can be grazed from May or June to September. Stocking rate of 4 to 6 steers per acre.

Zone B.

(1) Endophyte-free tall fescue, preferably with alfalfa, or white or red clover, can be grazed from late September to late December and from early March to June or July. Hay, silage, and corn, or similar quality feedstuffs can be fed in mid-winter to maintain gain at about 1 lb per day. Stocking rate of 1 to 2 steers per acre.

(2) Alfalfa can be rotationally grazed from April to July or August and again from September to early November. Stocking rate of 1 to 2 steers per acre.

(3) Endophyte-infected tall fescue can be used if at least 30 percent of the forage is a legume. Grazing period and stocking rate same as with noninfected tall fescue.

Zone C.

(1) Orchardgrass, with alfalfa, white or red clover, or birdsfoot trefoil, can be grazed from April to July or August. Stocking rate of 1.5 to 2 steers per acre.

(2) Endophyte-free tall fescue, preferably with alfalfa, white or red clover or birdsfoot trefoil, can be grazed from April to July and again from September to November. Stocking rate of 1.5 to 2 steers per acre.

(3) Endophyte-infected tall fescue can be used if at least 30 percent of the forage is a legume. Grazing period and stocking rate same as with noninfected tall fescue.

(4) Alfalfa, rotationally grazed from late April to September. Stocking rate of 2 to 4 steers per acre.

Zone D.

(1) Orchardgrass, with alfalfa, white or red clover, or birdsfoot trefoil, can be grazed from late April to September. Stocking rate of 1.5 to 2 steers per acre.

(2) Endophyte-free tall fescue, with white or red clover, or birdsfoot trefoil, can be grazed from late April to September or October. Stocking rate of 1.5 to 2 steers per acre.

(3) Endophyte-infected tall fescue can be used if at least 30 percent of the forage is a legume. Grazing period and stocking rate same as with noninfected tall fescue.

(4) Alfalfa, rotationally grazed from May to September. Stocking rate of 2 to 4 steers per acre.

Grain feeding of steers on high quality pasture during the cool season is generally not recommended as animals tend to substitute grain for forage. This allows higher stocking rates, but gain per day remains about the same unless an ionophore is added. If grain is to be fed, it is best to do this at the end of the grazing season when forage quality begins to decline. Light-weight steers, initially weighing 400 to 500 lb, are best suited for stockering as weight gains are high and more calves can be grazed on a given area of land. In some situations, stocker animals can be finished on pasture and slaughtered at about 1,000 lb for lean beef.

Summary

Stockering offers excellent opportunities for profit when operations are well managed. Several pasture systems are available in each climatic zone of the South. However, good pasture management is essential to maintain high forage quality throughout the season. Adjustment of stocking rate, supplemental feeding, and good health care are necessary to achieve the high gains necessary for profitability. ■

Chapter 28

Forages for Dairy Cattle

THE DAIRY INDUSTRY in the South has changed considerably over the years. Increased mechanization, improved breeding, better feeding, computerized ration balancing, advances in herd health, and better overall dairy farm management have resulted in more efficiency. Consequently, total milk production has increased dramatically in the last decade. This increase occurred with fewer cows on fewer farms. This trend, which is expected to continue, shows increased production per cow, increased herd size, and fewer farms. In the USA, there were approximately 21 million dairy cows on farms in the early 1950s. By 1994, this number had dropped to less than 10 million. During that period, herd size and production per cow had more than doubled.

Although the South has experienced growth in certain areas, a 10-year comparison shows overall changes in cow numbers similar to national changes. During the last decade, there was an 11 percent reduction in cow numbers but a 25 percent increase in production per cow.

Dairy producers have demonstrated their ability to efficiently produce quality products. Changing diets and lifestyles, availability of dairy substitute products, health concerns, and perhaps other factors have resulted in a decreased demand for some dairy products and a decrease in per capita consumption.

The dairy industry is responding with active promotion programs that bring the importance and value of dairy products to the attention of a consuming public. Likewise, the

Figure 28.1. Herd size and production per cow have increased sharply in the past decade.

industry is responding to consumer demands for different products such as low-fat milk and convenience products (sour cream and dips, yogurt, flavored milk). This provides evidence that the dairy industry is sensitive to needs and demands, visible, and progressive.

Feeding Systems

Use of pasture as a major provider of nutrients for dairy cows has varied considerably. In the early 1950s, pasture provided over 30 percent of the nutrients dairy cattle consumed. That value has fallen to less than 18 percent at present; however, interest in grazing is once again increasing for several reasons. Smaller dairies with sufficient land resources can make valuable use of high quality pasture. As herd size increases and land resources decrease, pastures often are used only for dry cows and replacement animals. Most large dairies do not have sufficient land for pasture, and have switched to a total dry lot feeding program. Pasture, dry lot, and a combination of the two systems have been used with success in research, demonstrations and on-farm experiences at several locations. The use of high quality pasture in conjunction with dry lot feeding can reduce feed costs and improve net profit from milk production on many farms.

The type of forage crops used for dairy pastures varies from one part of the South to another. In addition to being widely used for hay production, alfalfa is used to great advantage as a pasture crop by many dairy producers. In the upper South, orchardgrass-clover is a popular mixture. While endophyte-infected tall fescue is *not* a good choice, good milk production has been obtained from endophyte-free tall fescue. In the lower South, mixtures of winter annuals, especially ryegrass, are commonly used to provide quality grazing for dairy cattle.

High-Energy High-Quality Forage Necessary

High producing dairy cows require adequate levels of energy, protein, vitamins, minerals, and water in their diet. Of these components, energy and protein are of major importance because of the amounts required and the expense

Figure 28.2. High quality silage is readily consumed by dairy cattle.

Figure 28.3. Dairy cattle grazing high quality winter annual pasture near Dothan, Alabama.

of providing them. Feed cost is the largest single expense a dairy producer faces. High quality forages are the most economical source of the essential nutrients for dairy cattle.

High producing dairy cows must consume large quantities of high quality feed for optimum production. Dairy farmers are aware of the importance of forage quality. A high-quality forage is one that is palatable (consumed in large amounts), digestible (capable of supplying large amounts of nutrients), and contains the proper balance of needed nutrients. Forage quality can make the difference between high or low production and between profit and loss in dairy feeding programs.

Feeding high quality forage reduces feed cost, stimulates high milk production, increases dry matter intake, and contributes to health, thriftiness, and productive life of the cow. Conversely, feeding low-quality forage can be costly by reducing feed intake and milk production.

Energy. Although corn and other grains supply energy in dairy feeding programs, high-energy grain crops harvested and stored as silage provide the major energy source for most dairy operations in the South (see Chapter 20). Corn is rightfully considered "King of the Silage Crops." It has the potential of producing more energy per acre than other crops.

In addition to corn, sorghum and small grains are commonly used for silage. These crops produce good yields, are easily mechanized from standing crop to feeding, and are dependable.

Many factors influence silage yield and quality including: rainfall, variety, fertility, cultural practices, pests, stage of maturity when harvested, and harvesting and storing conditions. Corn and sorghum have the ability to produce large amounts of energy and nutrients per acre, but crude protein content is usually low and must be supplemented for high milk production.

Protein. Protein is needed for growth, production, reproduction, and maintenance. The amount required in the diet depends on animal size, condition, growth rate, and level of production. Protein content of forage plants varies considerably depending on many factors, especially species and stage of maturity. In general, legumes are higher in protein than grasses.

Legumes harvested at bud to early bloom, and grasses harvested at boot stage for first harvest and at a leafy stage for aftermath harvest, are usually high in protein. On dairy farms, alfalfa or an alfalfa-grass mixture is an excellent complement to corn silage. Other forages such as red clover grown in association with orchardgrass, endophyte-free tall fescue, or timothy can provide high quality dairy hay when harvested at the proper stage of maturity. Other cool-season species as well as warm-season species and various by-products can also be used in formulating dairy rations.

Forage Testing. In order to adequately and economically provide a balanced ration to the dairy cow it is important to know the nutrient content of feed sources to be used. With the wide variety of on- and off-farm feed sources throughout the South and the importance of forage quality to the dairy herd, forage testing is essential.

Forage testing provides the framework for ration balancing. Feed sources should be sampled so that representative samples can be submitted for analysis. Results from the analyses can then be used to provide a balanced ration without the extra cost of unnecessary supplements.

Grazing Opportunities. There is renewed interest in grazing by many dairy farmers as they adopt pasture-based systems. Grazing systems today are more adaptable and easier to implement primarily as a result of rapid advances in low cost fence and water systems (Chapter 24). The high degree of flexibility afforded by this technology allows a grazing system to be implemented quickly. It can be modified easily to minimize the adverse effects of seasonal forage quality and availability and to maximize utilization of forage resources to meet animal needs.

Other incentives for producers to shift to pasture systems include: increased soil conservation, fewer manure management problems, better animal health and reduced concern about various environmental issues. These are all important; however, a fundamental underlying issue is economics. With increasing cost of confined feeding systems and with little or no increase in milk prices, dairy producers are anxious to reduce production costs and improve profitability.

Feed comprises over 50 percent of the cost of producing dairy products in a traditional dairy operation. Grazing can mean a feed savings of 75 to over 150 dollars per cow per year. Grazing during only a portion of the grazing season can result in a one dollar savings for each one hundred pounds of milk. Seasonal grazing of good quality pasture can reduce corn silage usage as much as 80 percent.

There is no specific procedure required for success with grazing of dairy cows. The pasture contribution to overall feed production can vary considerably depending on resources, needs and desire. Some grazing enthusiasts have switched to a total grazing program with seasonal milk production matched to the main pasture season, while others utilize high-quality pastures when available to reduce a substantial amount of stored feed.

In general, high-quality pastures (grazing-tolerant alfalfa with adapted grasses, grass-clover mixtures, ryegrass, small grain) should be provided in a vegetative stage in adequate amounts for high producing dairy cows. This can generally be best managed with rotational stocking of small paddocks (Chapter 25). Many dairy producers use a forward grazing (Chapter 25) technique which allows the lactating cows first access to a paddock, followed by dry cows or replacement heifers. This represents an efficient utilization based on pasture quality and animal needs. Grain and mineral supplementation can be adjusted based on pasture supply, quality, and animal productivity.

Concerns often expressed by dairy producers considering grazing for the first time include bloat, fencing, water and cattle management. Although, these factors are of concern, experienced dairy producers do not consider bloat a major problem when proper precautions are taken (Chapter 21). Likewise, low cost, portable and flexible fencing and watering systems can be adopted and implemented to meet individual farm needs (Chapter 24 and 25). While limited work has been done in the South on the effect of cattle management on milk production, work in other countries has shown that cows can travel up to a mile from the parlor to pasture without a loss in production.

Summary

Dairy producers in the South have demonstrated their ability to produce quality products efficiently. Feed cost represents the largest single expense a dairy producer faces. High quality forages are the most economical sources of the essential nutrients needed by dairy cows. Knowing the quality of feed sources available through forage testing permits dairy producers to more efficiently and economically balance rations with off-farm feeds including by-products from industry.

Research and on-farm demonstrations have shown that pasture systems can be used to advantage by many dairy producers in the South. Survey results, demonstrations and research confirm that it is possible for experienced dairy producers with adequate resources to save $120/cow/year or more by adopting grazing programs to supply high quality pasture to meet at least some portion of the animals' feed requirements. The extent of such contributions depends on the quality and quantity of pasture available and length of the grazing season. Grazing high-quality, high-yielding forages such as alfalfa or alfalfa-cool season grass mixtures provides dairy producers an opportunity to reduce the amount of stored feed required and produce milk more economically. ■

Chapter 29

Forages for Horses

THERE ARE MANY REASONS why horse owners should have some knowledge of forage crops. Horses need some forage in their diet because it helps prevent digestive problems. An adequate quantity of good quality forage is also an economical way to minimize feed costs, which normally constitute the greatest expense involved in keeping horses.

In addition, lack of knowledge of certain potential problems which may occur in pasture situations or as a result of feeding an inappropriate type of hay, or grazing certain types of plants, may injure or even kill a horse. Horses are also less likely to chew on wood or other objects when in a good pasture. And finally, pastures provide a healthful exercise area for horses while beautifying the landscape.

General Nutritional and Feeding Information

Horses are natural forage eaters as evidenced by the fact that wild horses do quite well in open range situations. Feeding of concentrates becomes necessary only when there are great demands on the animals, such as when they are being worked or ridden a good deal. At other times, good quality pastures can totally meet the nutrient requirements of horses. If only low-quality forage is available, supplemental feeding may be necessary–especially for young, growing animals or lactating mares.

Horses are not ruminants. Their digestive systems differ considerably from those of cattle and sheep. Instead of having four stomachs,

Figure 29.1. Pastures such as this Kentucky bluegrass/white clover mixture in central Kentucky provide an exercise area and good nutrition while beautifying the landscape.

including a large rumen where fermentation and microbial digestion takes place, horses have a single, relatively small stomach. However, they do have an enlarged caecum near the end of their intestine in which bacteria digest forage somewhat like the rumen of a cow, although less efficiently.

The low digestion efficiency in the caecum results in a greater requirement for high-quality protein. Much of the protein obtained by cows and sheep is produced by bacteria in the rumen and subsequently digested in the ruminant's true stomach, or abomasum. Since the horse digests forage after materials pass through the true stomach, it is not able to digest microbial protein.

Because of the characteristics of their digestive system, horses need to be fed frequently. Furthermore, they need to consume good quality hay or pasture which will be easily digested and move rapidly through their digestive systems. How a horse is fed will greatly influence its growth, reproduction, speed and/or strength, as well as its appearance. As with any type of animal, the nutritional needs of horses vary depending on size, age, and activity. Diets should be formulated by qualified individuals based on the class of animal and the specific nutrient analyses of materials being fed.

The diets of horses can consist of a combination of pasture, hay, concentrates, and even high-quality, well-ensiled silage made from corn or small grains (however, silage should not comprise more than about 50 percent of the roughage intake). Once a diet has been formulated, abrupt shifts in feedstuffs should be avoided, especially from very high quality to very low quality. Gradual changes using mixtures are much less likely to cause digestive upsets.

Hay for Horses

Hay is an important feed for horses. Non-lactating broodmares and idle mature horses can do well on high-quality hay alone, while increased activity and reproduction require supplemental feeding. Even when other supplemental feed is being provided, it is desirable to furnish horses with hay because it tends to prevent colic and digestive problems. Horses need to receive about 1 to 2 lb of high quality hay per 100 lb of body weight.

Hay and other feed materials provided to horses should be free of mold; otherwise, digestion may be upset. Dry, dusty hay can also irritate the lungs, resulting in a cough, runny eyes or nose, and possibly pneumonia. Good horse hay should be free of weeds and foreign objects. Hay should also be leafy, and have a pleasant aroma. Fine-stemmed, soft hay is preferable to coarse, fibrous hay.

A misconception is that good hay will always be a bright green color. Although a bright color is an indication of no rain damage, color alone is a poor way to evaluate hay because low quality hay can have bright color while top quality hay will eventually lose color with time. Horses are color blind, so hay color makes no difference to them.

When purchasing hay, it is best to buy it by the ton if possible because it is considerably cheaper when bought in quantity. Also, when purchased by the ton, one does not have to be concerned with variability in bale size, which greatly affects the value of hay purchased by the bale.

Although visual inspection is quite useful, the best way to evaluate hay quality is by a laboratory analysis. Many reputable hay dealers have their hay tested as a routine practice. Beware of any hay dealers who do not want an analysis of their hay!

"Hay belly" is a term which some horse producers use to refer to animals having a large, distended digestive tract. This occurs as a result of feeding large quantities of low-quality roughage. High-quality hays do not cause this problem.

Pasture for Horses

Providing pastures for horses is a healthful way to allow them to get exercise while simultaneously furnishing the nutrients they require. The amount of pasture required for a horse will vary depending on factors such as size and age of the horse, pasture species, amount of supplemental feed provided, and soil fertility. However, for most of the South, one or two acres of pasture should be provided for each mature animal.

Mature, non-working horses and well-developed older yearlings can normally be maintained on good pasture with little or no grain supplementation. Animals with higher nutritional requirements, such as young growing horses or work horses, will need some grain and high quality hay along with pasture, as previously discussed.

Many different forage plants can be used for pasture in the South, and most make suitable pastures for horses. While horses will consume some forage of most legumes, they generally prefer grasses. In most cases, horse pastures should consist of grasses or of grass/legume mixtures. However, arrowleaf clover and vetch are two annual legumes which are quite unpalatable to horses.

Bermudagrass, bahiagrass, and dallisgrass are often used to pasture horses in the lower South. As discussed in Chapter 12, these species

Figure 29.2. In the lower South, bermudagrass or bahiagrass makes good summer pastures for horses.

can be overseeded with winter annuals to extend the grazing season.

In the upper South, Kentucky bluegrass and orchardgrass are commonly used for horse pastures. In addition, endophyte-free tall fescue makes an excellent horse pasture, but in the interest of long-term persistence, care should be taken to avoid overgrazing. As discussed elsewhere in this book, white clover, red clover, or birdsfoot trefoil can be used as companion species with these cool season grasses.

Winter annual grasses (small grain and/or ryegrass) or winter annual grass/legume mixtures make superb horse pastures. Summer annuals suitable for horses include pearl millet, browntop millet, and crabgrass. With management to allow reseeding, crabgrass can be a dependable volunteer species following winter annuals planted in autumn on a tilled seedbed.

Browntop millet makes a good quality, but short-lived pasture. Pearl millet is much more productive, but may be initially unpalatable to the animals. Due to its erratic growth, pearl millet is also difficult to keep stocked properly.

Species to Avoid

Sorghum, sudangrass, Johnsongrass, and particularly sorghum-sudangrass hybrids are **not** recommended for horse pastures. Horses grazing these species may develop cystitis, which results in paralysis and urinary disorders. Hay of these species is safe to feed to horses, however.

Endophyte-infected tall fescue should **not** be used as pasture for pregnant broodmares, especially during the last three months of gestation (see Chapter 23). Endophyte-infected tall fescue has also been shown to reduce weight gains of yearling horses and may have other effects which have not been documented. Non-infected tall fescue makes an excellent horse pasture or can provide good quality horse hay.

Figure 29.3. Winter annuals make excellent horse pastures.

Figure 29.4. Mares often have reproductive problems on endophyte-infected tall fescue, but non-infected tall fescue pasture or hay is excellent for horses.

Special Pasture Management Considerations

Horses are more selective than cattle and tend to "spot" graze. This is particularly true if legumes are present in the pasture. Horses will eat legumes, but tend to prefer grasses. Therefore, horse pastures may require periodic clipping to even-out the overgrazed and undergrazed areas and to keep forage plants young and tender.

If pastures are overstocked, horses will not have the opportunity to selectively graze, and nutritional intake will drop. Grazing horses and cattle together is desirable because cattle reduce the effects of spot grazing. Also, horses and cattle will graze around each other's fecal and urine spots, but not their own.

Pastures for horses should be well drained, free of holes or stumps which might cause injury to the animals, and contain no poisonous plants (see Chapter 22). It is important to have safe fences which will not cut the horses. Fertilization should be according to soil test recommendations. Horses need access to shade, plenty of clean fresh water, salt, and a basic mineral mixture. It is desirable to graze mature horses separately from yearlings. Horses do not normally bloat or founder on pasture but may scour on an extremely lush pasture.

Spraying of horse pastures with an appropriate herbicide to control weeds is highly desirable when weeds are a problem. **Any grazing restrictions provided on the herbicide label should be strictly observed.**

The hooves of horses can be more damaging to pasture than those of cattle. Thus, rotational stocking is desirable, particularly in newly established pastures. It may also be necessary to protect even a well-established pasture by temporarily removing the animals during wet periods when the ground is soft.

Parasites can be a problem where horses are being pastured. This is particularly true when pastures are heavily overgrazed. Periodic chain harrowing of pastures to scatter fecal matter can reduce parasites, except during wet weather.

Summary

The nutritional requirements, grazing habits, and forage preferences of horses differ from those of cattle. In addition, horse producers need to be aware of cautions pertaining to a few forage crops. Nonetheless, most forage species adapted in the South make quite acceptable pastures, or provide good quality hay, for horses. With

Figure 29.5. Kentucky bluegrass is a commonly used forage in horse pastures in the upper South.

the investment of only a little time and study, Southern horse producers can plan a forage program which will safely provide a nutritious, relatively inexpensive diet for their animals. ■

Chapter 30

Forages for Sheep

SHEEP lead all farm animals in the ability to produce marketable products on forage alone. Therefore, for persons who have developed the philosophy that their commodity is forage and animals are the machines they use to harvest it, sheep production is an enterprise worth considering.

Harvesting Forage with Sheep

Except for certain highly fibrous summer grasses, most improved pasture species normally used in the South provide adequate nutrition for sheep (see Chapter 17). Grains or other concentrates generally are needed only in late pregnancy and early lactation to supplement low-quality forage, or for rapidly growing lambs when forage quality is not sufficiently high. A ewe nursing twins will require more nutrients than a ewe with a single lamb. As is the case with cattle, energy is more likely than protein to be a limiting factor in sheep production.

The economic outlook for sheep production from pasture appears favorable, if good management is used. Successful sheep producers must give careful attention to parasite and predator control, and to births during lambing season. Sheep may be pastured alone or may be grazed with cattle. In fact, sheep and cattle complement each other since they have different grazing preferences.

Sheep are highly selective grazers and prefer short, fine forages, while cattle take larger bites and are less selective. Despite their higher selectivity, sheep tend to graze broadleaved plants, including many weeds, to a greater extent than cattle. Thus, a pasture containing both cattle and sheep, or grazed alternately by cattle and sheep, is more likely to be better utilized and require less mowing. Another advantage of grazing both sheep and cattle is that parasite levels for both types of livestock tend to be reduced.

A very high percentage of the annual feed requirements for sheep can be provided from forage. A profitable sheep operation depends on providing as much grazing throughout the year as possible. Although the stocking rate for sheep varies according to forage species and soil fertility, usually about 4 to 6 ewes can be grazed on the same amount of pasture required for 1 cow.

Usually, 30 to 35 ewes plus 1 ram is considered to be the minimum economic size for a sheep flock. Some sheep breeds are better suited than others to the climate of the South. These include Suffolk, Dorset, and Hampshire.

Matching Breeding Season to Feed Supply

In a sheep enterprise it is very important to match the breeding cycle to the availability of high quality forage. The best time for lambing may vary considerably within the region, depending on the forages available at various times of the year.

Late pregnancy and early lactation are the times when ewes have the greatest energy requirements. Thus, in most of the South, a late

Figure 30.1. Winter annuals make excellent pastures for sheep.

winter or early spring lambing program is generally most successful. In such a program, breeding would occur in the autumn and the pregnant ewes would be wintered on hay, silage, or cool season pastures. With a program of this type, the lambs can either be grazed through the spring and early summer on high-quality pastures, or they can be weaned at 90 days of age or less and fed out in a drylot.

Autumn lambing is a possibility, but is less likely to be profitable due to a lack of nutritious autumn forage. Such a program may be successful when alfalfa or cool season annual grasses are utilized extensively. In such a program the ewes would be bred in late spring and the lambs marketed in mid-spring after having grazed winter annual pastures, orchardgrass/white clover, or endophyte-free tall fescue/white clover pastures. Pasture systems in the various climatic zones will be similar to those for stocker cattle (see Chapter 27).

Immediately after lambs are weaned, the nutritional requirements of a ewe is low. Thus, at this time ewes can be maintained on low quality forages until near the beginning of the breeding season.

"Flushing" is a term which refers to the practice of increasing nutrition levels of ewes just prior to breeding. This is usually done beginning about 3 weeks prior to the breeding season, and continued until about a month after conception. It has the effect of increasing ovulation rates as well as the percent of twinning and therefore is highly recommended. It can be done either by providing the animals with high quality pasture or by supplemental grain feeding.

Internal parasites can be a severe problem for sheep, and efforts should be made to prevent or overcome this problem. Periodic worming, perhaps as often as monthly, is essential in the South. In addition, parasite levels can be kept lower by not keeping sheep on the same closely-grazed pastures for a long period of time.

It is necessary to provide minerals for sheep. Most sheep producers do this by keeping trace-mineralized salt available for their animals at all times.

Wintering of Sheep

Sheep can be wintered on pasture, hay, or silage. In Zones A and B, cool season annual pastures provide high quality forage and have a relatively short period of winter dormancy.

Most sheep producers find it necessary to feed hay during certain portions of the year. Legume hays such as alfalfa and red clover are particularly useful, but good grass hays are also acceptable if harvested at an early stage of maturity.

Corn or legume silage can be an acceptable winter feed for sheep if properly supplemented with protein and minerals as needed. However, most sheep producers prefer to feed some good quality hay along with silage.

The problems associated with growing sheep in the South are well defined. Predators, especially dogs, have been a serious problem in many areas, but the advent of high voltage electric fencing has greatly reduced this problem. Donkeys will bond well with sheep and can also provide predator control. In addition, parasites and diseases which tend to be more prevalent in the South must be controlled. A veterinarian or livestock specialist can provide advice for dealing with these problems.

Finally, the market for meat lambs, as well as for wool, in a particular area should be carefully evaluated before initiating a sheep operation. Markets for feeder lambs are much more limited than for feeder cattle. Many sheep producers have had no particular problem with marketing, but this is an area worthy of careful consideration prior to investment of a great deal of resources.

Pastures for Lambs

Special consideration must be given to the nutrition of lambs. If high quality pastures are not available for the entire flock, it may be advisable to creep feed or creep graze lambs in order to provide them with the higher levels of nutrition which they require. If neither of these options is feasible, then confinement feeding is desirable.

In some situations, producers do not keep ewes, but rather buy lambs which are grazed on high quality pasture. Orchardgrass and white or red clover, endophyte-free tall fescue and clover, or alfalfa are good choices for lamb grazing, but precautions should be taken to prevent bloat. Successful grazing of lambs requires a high level of pasture management and avoiding accumulation of surplus forage of decreasing nutritive value. Feeding grain along with pasture may increase gains and will certainly allow an increase in stocking rates, but may not be economical.

Summary

As compared to the rest of the USA, the South has relatively high rainfall and a long growing season. This situation is conducive to the growth of many different forage crops. In the South, sheep can be grazed over much of the year. This constitutes an opportunity for producers interested in raising sheep.

Sheep production allows conversion of a higher percentage of the forage produced into marketable products than does any other Southern livestock enterprise. Since steady progress has been made in overcoming problems, it appears that sheep production is a logical enterprise in the Southern USA for producers who are willing and able to exercise the management required. ■

Chapter 31

Plants for Wildlife

MANY RESIDENTS of the South are interested in wildlife. Hunting is a popular sport enjoyed by millions, and many non-hunters enjoy observing wildlife and helping provide better conditions for wild animals.

Before European settlers arrived on the North American continent, wild game was plentiful. However, as land was cleared and farming practices led to the destruction of wildlife habitat, populations of many wild animals declined. Not surprisingly, many Southerners would like to see wildlife populations increase.

Through the years, many approaches to increasing wildlife numbers have been tried. These have included the release of pen-raised animals into the wild, imposing of strict regulations pertaining to hunting, and predator control. Depending on the situation, these techniques may or may not be helpful, but the best method of increasing wildlife populations is to improve the habitat by: (1) increasing the amount of cover; and/or (2) improving the year-around food supply available for the animals. Many forage crops can accomplish these purposes.

Characteristics of Game Animals

Before discussing plants which are useful in feeding or protecting wildlife, it is worthwhile to first consider a few of the habits and characteristics of common game animals. The behavior patterns of animals have a major influence on the appropriateness of growing

Courtesy of Neil Waer

Figure 31.1. Fawn in white clover near Auburn, Alabama.

various crops, as well as how they should be planted.

Deer. The range of white-tail deer can vary widely, but often is an area approximately three-quarter mile wide and one and one-half miles long. Usually, a buck will range over about a section (640 acres) of land, and a doe will range over about one-half that. However, deer will travel substantial distances from their normal range to reach a food plot. Also, bucks range widely during the rutting season. Thus, it is quite possible to attract deer to an area for the purpose of harvesting them.

There has been relatively little research done to investigate deer preferences or the value of various plants in deer nutrition. However, winter is generally the time when food is least available and it is known that winter annuals such as rye, oats, wheat, ryegrass, and crimson clover are good choices.

High-quality perennials such as white clover or alfalfa are also excellent. There is little known about deer preferences for summer plants or their effect on deer growth and development.

Doves. Although some mourning doves (commonly called simply "doves" in the region) stay in much of the South all year, the dove is basically a migratory animal. Millions of doves fly through the South each autumn and spring. In addition, even "local" doves may range over a wide area in search of food. Non-migrating doves will fly up to 12 miles to reach a feeding field.

Doves have a high mortality rate, with an estimated 70 percent of the population dying from one year to the next. Although it may be important to a local population, hunting pressure has relatively little effect on dove mortality overall. The high rate of mortality is of little consequence because doves are extremely prolific, often raising four to six broods per year. In addition, some doves hatched early in the year can reproduce before the year end.

Doves can be attracted with food plots, but it may be difficult to hold them in a given area. Their migratory nature and lack of hesitation to abruptly abandon one feeding area for another account for the unfortunate experience many hunters have had of seeing large numbers of doves in a field one day, but none the next when a dove shoot has been arranged. Such rapid movement is often accompanied by a change in the weather.

Probably the best plant species to attract doves are browntop millet, proso millet, and sunflower. A small food patch of two or more acres should be planted in early spring to feed local doves, with a larger planting in mid-to-late spring to attract migratory doves.

Doves are poor scratchers and will not dig up buried seed. Also, they usually will not perch on plants to eat from seed heads. They prefer their food lying out in the open on relatively bare ground.

Ducks. With the exception of some wood ducks which remain in the South throughout the year, ducks are migratory birds. Food plots are commonly used to attract migrating ducks. In flooding areas, Japanese millet, corn, and grain sorghum are good duck food plants. In other areas, browntop millet is an excellent choice.

Quail. There are several species of quail native to the USA, but in most of the South, the most common quail is the "bobwhite." Other species of quail have similar feeding habits to the bobwhite.

Quail covey ranges are usually fairly small. The size or shape of a range varies, but is usually less than one-half mile in width or length. The requirements of a range are: (1) adequate food availability, (2) roosting sites, (3) nesting cover, (4) escape cover, and (5) protection from bad weather. All of these except nesting cover must be available throughout the year.

Quail eat a diversity of food items. Their diet consists mainly of seeds, fruits, and insects, but they will also eat small amounts of green leaves. Probably the best plants to grow for quail food are annual lespedeza, partridge pea, shrub lespedeza, reseeding vetch, and Egyptian wheat. Millets, corn, sorghum, and alfalfa are also excellent quail plants.

The bobwhite seems to thrive best where there are about equal amounts of cultivated crops, idle fields which have not been under cultivation for 3 to 10 years, woodlands, and areas reverting to woodland. It is best if these are in small, well-scattered tracts.

One acre of food plots for every 12 acres of land usually provides adequate food for high quail populations. Food plots should be at least 15 feet wide and at least one-quarter acre in size. Cover plots should be at least one-eighth acre in size. Several one-eighth acre plots are better than one large area. Annual (striate or Korean) lespedeza is probably the single most important winter food of quail in the South.

Rabbits. Rabbit populations are dynamic, and food supplies are critically important in maintaining a high population. Since winter is the time when rabbit food supplies are of most concern, any type of green winter plants (with the exception of endophyte-infected tall fescue) is a good choice for rabbit food plots.

When food plots are planted for rabbits, it is best to plant them in narrow strips (perhaps 10 feet wide) close to a thicket or other area which provides cover. This allows the rabbits to feed, but they still have the opportunity to scamper back into the cover if danger threatens. Examples of good plants to grow for rabbits include various clovers, small grain, and ryegrass. Deer and wild turkeys will also use plantings of these species.

Wild Turkeys. Turkeys range over a wider area than deer and eat a varied diet consisting of insects, seeds of various plants, and a substantial amount of green leaves. Chufa and any type or mixture of winter annuals are good choices for turkey food plots.

Plants for Wildlife

Plants provide both cover and food for wildlife, but this discussion will center primarily on plants to provide food. Although most of the plants discussed are forage crops, a few non-forage species are included.

When considering plants for wildlife food, the first question which comes to mind is, "What will wildlife eat?" The answer is that wildlife will, if forced to, eat many different things. Dramatic evidence of this is provided by areas which have excessively high deer populations. During periods when there is little plant growth, virtually every bit of green material up to the maximum height which deer can reach will be consumed.

The primary concerns relating to plant selection for wildlife food plots relate to species palatability to wildlife, nutritive quality, quantity of food produced, and length of time of food availability, especially during periods of critical need. Plants which meet these requirements should be considered for wildlife plantings. The following is a discussion of some plants which have been found to be useful in wildlife plantings or beneficial to wildlife on a farm if planted primarily for other purposes.

Alfalfa. This cool season perennial legume makes excellent high quality forage growth during spring, summer, and autumn. The forage is relished by deer, and it is an excellent source of insects and green material for quail and turkeys. While much more management-intensive than most wildlife plants (see Chapter 8), it has a great deal to offer in certain situations. In many cases it may not be feasible to plant alfalfa specifically for wildlife, but the crop certainly can be beneficial to wildlife. The availability of grazing-tolerant varieties has increased the feasibility of planting the crop in areas where wildlife populations are high.

American Jointvetch. American jointvetch is also referred to as Aeschynomene. This plant is not a vetch, nor is it a cool season forage as are other vetches. American jointvetch is a summer annual legume which is sometimes planted for deer. It is best adapted to fairly moist sites in Zone A (**Figure 3.2**), but can often be successfully grown on well-drained soil. There have been mixed reports regarding its value as a food source for deer.

Annual Lespedeza. Annual lespedeza includes both Korean lespedeza, of which there are several varieties, and striate lespedeza. Korean lespedeza is best adapted in the upper part of Zones B and in Zones C and D, while striate lespedeza is the best choice for the lower part of Zone B and for Zone A.

Annual lespedeza is an extremely useful species to plant for quail food. Doves are not particularly fond of annual lespedeza, but will eat it if nothing else is available.

Browntop Millet. Browntop millet is an extremely valuable plant for attracting doves, quail, or ducks. Timing of planting can be an important consideration with browntop millet because seed mature about 60 days after germination. Consequently, if browntop is planted too early, all the seed may be gone before dove season begins. Therefore, a small planting for local doves should be made in early spring, with a later planting aimed at attracting the much larger numbers of migratory doves in late season. It is best to time a planting so that doves have at least 2 weeks of feeding time to become accustomed to the field before a shoot is held. All plantings need to be made early enough to allow the plants to mature seed before frost.

Chufa. Chufa is a type of giant nutsedge which is a favorite food of deer and especially wild turkeys. The chufa is widely adapted in the South and is easy to grow. It is best suited for use on soils which are sandy or loamy.

Wildlife do not eat chufa leaves, but they relish the nutlets which grow on chufa roots. When the plants are mature, a few plants should be dug to expose the nutlets. Once wildlife have discovered the nutlets, they will dig the rest themselves. Chufa is planted primarily for turkeys. It probably is not worthwhile to plant them just for deer.

Although chufa will sometimes volunteer in an area in subsequent years, it is not as aggressive as many other types of nutsedge. Therefore, chufa is not likely to be a problem in subsequent years in a field where other crops may be planted.

Egyptian Wheat. Egyptian wheat is a type of grain sorghum. It is not an outstanding grain yielder, but it

Courtesy of National Wild Turkey Federation/John Gwaltney

Figure 31.2. Wild turkeys range over a wide area. Winter annuals are good choices for food plots.

has loose heads, which allows birds easy access to the grain.

Egyptian wheat is an especially good choice for quail. The seed shatter over a fairly long period of time, thus providing an extended period of food availability. Furthermore, the 6- to 10-foot stalks tend to lodge easily, therefore providing cover for quail while they are feeding. This gives the birds a sense of security and protection from avian predators. Also, as compared to some food plants used for quail, deer are not likely to consume large quantities of Egyptian wheat.

Florida Beggarweed. Florida beggarweed is always in demand as a quail feed. Because this plant has a long growing season, and may be killed by frost prior to making seed if grown too far north, its usefulness is limited to Zone A. It is best adapted to fertile, moist, sandy soils. Seed of this plant need to be scarified before planting and commercially-available seed usually will have been scarified. Several species of beggarweed are native to the South.

Foxtail Millet. Foxtail millet is infrequently planted but also makes good food for birds. This species is best adapted in Zones B and C. Foxtail millet normally produces mature seed in about 75 days after germination.

Japanese Millet. Japanese millet can be grown for all game birds, but it is especially well suited for ducks. It can be grown successfully on well-drained soil in all zones, but it can also tolerate flooded soil as long as a part of the plant is protruding from the water. Most varieties mature within 80 to 90 days, but it is possible to use varieties which mature in 120 days.

Other Sorghums. Although Egyptian wheat is one type of sorghum commonly used for wildlife plantings, grain sorghum is also often used in wildlife food plots. Most modern sorghum hybrids will provide excellent high energy food for quail and doves. If sorghum is planted over a large area, strips should be mowed through the food patches at one-month intervals during the fall and winter to give the birds access to the grain. Sorghum is quite sensitive to soil acidity. If the pH is lower than 5.6, lime should be applied and worked into the soil before planting.

Cowpea. Cowpea, also called southern pea, is well-adapted to many areas in the South. This plant makes a substantial amount of summer growth which can provide forage for deer, cover for quail, as well as seed which can be consumed by various types of birds. If all the seed are not consumed, cowpea has the potential to reseed. Reseeding can be facilitated by disking a seed-containing area in the spring before cowpea germinates.

Several varieties of cowpea are available, including some types selected specifically for wildlife purposes. The plant is easy to grow, soil pH is not critical, and the crop does well with minimal fertility. However, the seed should be inoculated with cowpea inoculant prior to planting.

Partridge Pea. Partridge pea is one of the most widely planted quail food plants. It is tolerant of a wide range of soil types and soil fertility conditions, but is especially well adapted to clay soils in Zones A and B. Seed of this summer annual legume are usually both in demand and expensive.

Partridge pea produces a large quantity of hard seed, and once a stand has been established in an area there will usually be a good deal of reseeding year after year. This is especially true if the soil is disturbed in late spring. When partridge pea is first established, scarified seed should be used. In order to obtain reseeding, the area should be disked or burned in late winter or early spring. Because of its hard, durable seed, partridge pea provides food for quail over a long period of time.

Proso Millet. This is another summer annual grass which is quite attractive to doves and quail. There are both white proso and red proso millet types, with the white proso being especially useful. Like browntop millet, proso millet is a quick-maturing crop, but it has the potential of producing a higher seed yield than browntop.

Sericea Lespedeza. Quail will eat sericea lespedeza seed, but its primary value for quail is as a cover. Deer will readily eat the forage of low-tannin sericea lespedeza but are not fond of the high-tannin varieties. Sericea lespedeza is rather infrequently used in wildlife plots, but is often of benefit to wildlife when planted for livestock pasture or hay.

Sesame. Sesame or benne is highly palatable to both quail and doves. Most varieties shatter very quickly. The seed produced is available in great quantities during September through November, but they are soft and deteriorate quickly. This species is widely adapted throughout the South. It is an excellent source of feed for both game and non-game birds.

Sesbania. Sesbania is an excellent quail food plant which can also be used by doves. Once a stand has been obtained in an area it is likely to volunteer in subsequent years (perhaps being a serious weed problem if the area is row cropped). It is best adapted to wet, clay soils in Zone A. The seed are not highly palatable, but it is an important food source because the seed remain in good condition until late winter when a food shortage often exists. Sesbania should be planted in early spring.

Shrub Lespedeza. Bicolor lespedeza is often planted along field borders or in small patches to attract and feed quail. A related shrub lespedeza is Thunberg lespedeza (*Lespedeza thunbergii*). Both of these are long-lived perennial shrubs which produce large quantities of seed. Thunberg lespedeza is recommended in areas having high deer populations, since deer are less likely to consume this species.

Shrub lespedeza can grow so thick that it hinders dog handling and shooting. Consequently, it is best to plant it in strips between fields or along the borders of woods. Both species can be established from either seed or seedlings, but survival and establishment are usually best with seedlings. Seed should be scarified.

Many people have found that if they have patches of shrub lespedeza, it tends to help them locate quail easily. Quail are fond of the seed, and also feel protected from avian predators by the upper portions of the plant. Therefore, during the winter they often tend to spend most of their time in patches of shrub lespedeza. Thus, a hunter can move from patch to patch with a fairly high probability of finding many of the coveys in an area.

Shrub lespedezas are best suited for soils of medium texture and good drainage. Every 3 to 5 years the plants should be mowed to a 4- to 8-inch height in late winter in order to thicken the stand and prevent the plants from becoming too tall.

Sunflower. Sunflower is highly attractive to doves and other game birds. Several small-seeded, high oil varieties have been developed primarily for wildlife purposes, but any sunflower variety is acceptable. The usual length of time to maturity for sunflower is about 100 days. Sunflower is pollinated by bees, so a good bee population is necessary to obtain good seed set.

Velvetbean. Many years ago velvetbean was commonly planted with corn as a livestock feed in the lower South. Corn was harvested by hand before allowing livestock access to eat the fodder and velvetbean. In recent years, there has been a revival of interest in growing velvetbean and corn as a summer deer food. Although this should normally not be a high priority crop for deer, it is an option.

Winter Forage Crops. Various winter annual forage species including mixtures of wheat, rye, oats, annual ryegrass, and/or any of a number of winter annual clovers make excellent "green fields" for deer. White clover and alfalfa can also be used in a similar manner. In addition, such fields provide cover, seed, and a source of insects for both game and non-game wildlife species.

All species of vetch produce seed which make good bird food. However, this seed is mostly available in summer when there is usually plenty of other food available. Hairy vetch is adapted throughout the South, but several varieties of common vetch are adapted only in Zone A. An excellent use of vetch as a wildlife plant is "green feed" patches to provide winter green material for deer or turkey.

Reseeding common vetch is also an excellent plant to attract insects for feeding by turkeys or quail in early spring. Since quail need high protein insects for their breeding season, vetch can thus play a vital role in reproduction.

Wildlife Plant Establishment

There are limited data available on the amount of money and energy spent on establishing wildlife plantings in the South, but it is certainly many millions of dollars annually. It is probable that the greatest amount of money is spent on plantings for deer, followed by quail, and then doves. However, interest in establishing plants to provide food and cover for wildlife is increasing for all game wildlife species.

As with any crop (see Chapter 11), it is important to follow recommended planting procedures when establishing a wildlife crop. Otherwise, results are likely to be disappointing. Fertilization and liming should be according to soil test recommendations. General guidelines regarding planting dates, seeding rates, and seeding depths for plants discussed in this chapter are provided in **Table 31.1** and **Table 11.1.** Wildlife plant characteristics are given in **Appendix A.32**.

Summary

It is likely that there will be continued and even increased interest in planting food and cover plots for wildlife in the South in the future. Many people have the financial resources to join hunting clubs or privately leased land for hunting. In addition, the financial situation of many landowners is causing increased interest in alternative enterprises including catering to hunters. Further, we are learning more about how wildlife can benefit from establishment of various types of plants.

Wildlife is already a big business in many parts of the South and will continue to grow in importance. It is likely that many forage crops, as well as other crops, will be used in ever-increasing quantities for wildlife purposes. ■

Table 31.1. Seeding rates and usual planting dates (Zone A used for example) for species commonly planted for wildlife purposes.[1]

Plant		Seeding rate[2] lb/A	Planting dates
American jointvetch	B:	20-30	4/1-6/30
Annual lespedeza	B:	25-35	2/15-3/31
Beggarweed	D:	7-8	2/15-3/31
	B:	12-16	
Browntop millet	D:	15-20	4/1-8/15
	B:	25-30	
Cowpea	D:	30-40	5/1-6/15
	B:	100-120	
Chufa	B:	30-50	4/1-6/30
Proso millet	D:	15-20	4/1-8/15
	B:	25-30	
Egyptian wheat	D:	5-8	4/1-5/30
	B:	15-20	
Foxtail millet	D:	12-15	4/1-8/15
	B:	25-30	
Grain sorghum	D:	5-8	After frost
	B:	15-20	
Japanese millet	D:	15-20	4/1-8/15
	B:	25-30	
Partridge pea	D:	7-8	2/15-3/31
	B:	12-16	
Sesame	D:	8-10	4/1-5/30
	B:	10-12	
Sesbania	B:	15-18	3/15-5/1
Shrub lespedeza	D:	8-10	3/1-4/15
	B:	15-20	
Sunflower	D:	15-20	4/1-6/30
	B:	35-40	
Velvetbean	D:	15-20	3/15-5/15
	B:	50-60	
(often planted with low seeding rate of corn)			

[1]Seeding rates for other forage plants commonly used for domestic livestock but which may also be planted for wildlife are provided in Chapters 5-8.
[2]D = Drilled; B = Broadcast

Chapter 32

Forages and the Environment

IN RECENT YEARS there has been increasing awareness of the importance of preserving our natural resources and the environment. It appears that increased emphasis on environmental preservation and protection is likely in the future.

Agriculture has been one of the targets of criticism regarding environmental matters. Some people think agriculture has become too intensive, wasteful, and reckless and that radical changes need to be made in production practices. On the other hand, some agriculturists warn that changes will threaten our food supply and the economic viability of agricultural production.

While the environmental debate continues, it is certain that our "environmental awareness" level has been raised. It now seems increasingly desirable to assess the impact agricultural production is having on the environment. Fortunately, forage agriculture makes numerous **positive** contributions, as discussed in this chapter.

Soil Conservation

It has long been recognized that preserving the soil is extremely important to individual farmers and landowners as well as to our nation. Concern about this problem, heightened by tragic abuses, led to the creation of the Natural Resources Conservation Service, a federal agency which has been remarkably effective in reducing soil erosion problems. Yet, wind and water erosion remains a serious problem in the South.

Soil erosion causes numerous damaging effects. Streams, rivers, ponds, and lakes become polluted with soil sediment, adversely affecting fish and other marine life, as well as water use by humans. When topsoil is lost, the remaining soil is usually lower in organic matter and has a lower capacity for holding nutrients and water. Nutrients, both inherent and applied, are lost along with soil and may also contribute to lowered water quality. Gullies and ruts are created which make the land more difficult to use for agricultural purposes, and the remaining eroded land is certainly less productive and less valuable.

Clearly, the human race must take responsibility for soil erosion, because nature endeavors to avoid it. If simply left alone with no interference from man, almost any area in the South will soon develop a vegetative cover which protects the soil from erosion. Unfortunately, while nature will protect the soil, it does not generate food and profits to the extent which humans need and expect.

The solution to this dilemma is to use proper tillage methods on non-erodible areas for crop production and to use more erodible areas for other purposes, allowing economic production without risking excessive erosion losses. Most farms in the South have significant acreages which should not be tilled, or be tilled only infrequently. For millions of acres of highly erodible land in the region, growing forage crops is an alternative to row crop production.

Perennial grass sods are particularly effective in greatly reducing soil erosion losses (**Table 32.1**). Even if a given field is used for row crop production, forages may be used in conjunction to reduce erosion. Grassed waterways, grassed terraces, strip cropping, and long-term rotations utilizing forages are good examples. Many people do not fully realize the potential damage from unchecked erosion. Once serious erosion damage has occurred, there is little which can be done to restore productivity.

Table 32.1. Estimated average annual sheet and rill erosion on non-federal rural land in selected states.[1]

State	Erosion loss (tons/A)	
	Cropland	Pastureland
Alabama	6.2	0.6
Arkansas	3.4	1.2
Florida	1.2	0.1
Georgia	5.1	0.4
Kentucky	5.3	3.2
Louisiana	3.5	0.2
Mississippi	5.5	1.2
North Carolina	5.3	1.0
Oklahoma	2.8	0.7
South Carolina	3.2	0.4
Tennessee	7.1	0.7
Texas	2.5	0.5
Virginia	4.6	3.6

[1]Source: Summary Report, 1992 National Resources Inventory, USDA-Natural Resources Conservation Service (Issued July 1994, Revised January 1995).

Stabilization Of Critical Areas

The use of forages to prevent erosion is certainly not limited to farms. Disturbed areas often need vegetation to protect the soil and beautify the landscape. When this is the case, it is likely that forage crops will be used to accomplish the job. The result is that millions of acres of nonfarm land are planted to forage crops.

Industrial construction sites, roadbanks, parks, landfills, mine reclamation sites and even portions

Courtesy of Morris Gillespie

Figure 32.1. Strip cropping with forages helps protect water resources from pollution.

of golf courses are often seeded to forage crops. Initially, these sites are likely to be devoid of topsoil, inherently low in soil fertility, and often have high soil acidity. Mixtures of annual and perennial grasses and legumes are often seeded together to quickly stabilize the soil until the best adapted perennial forages have an opportunity to dominate.

With time and proper management, most disturbed or critical areas can be stabilized. Eventually, the soil is improved with the addition of organic matter, and a former eyesore may become an attractive area due to the beneficial effects provided by forage crops.

Water Quality

Improvement of water quality has become one of the nation's top priorities. Without water, there would be no life. Contamination of water jeopardizes all life.

In most areas of the South, the primary contaminant of water is soil particulate matter. Growing forage crops greatly reduces erosion as compared to other agricultural land uses. If the percentage of cropland devoted to forage crops was substantially increased, there would be a considerable improvement in overall water quality.

The other primary concern regarding water quality is contamination with certain inorganic compounds, especially nitrates. There have been cases in which nitrates have contaminated ground water, but this is relatively uncommon. Over-fertilization is sometimes perceived as being a common agricultural practice, but under-fertilization is more likely to be the case. Farmers cannot afford to waste money by purchasing more fertilizer than is needed and, in fact, many croplands are deficient in the primary nutrients.

There is now increased awareness of potential environmental contamination due to nitrogen (N) fertilization associated with intensive row crop production or to concentration of large numbers of animals such as

Figure 32.2. Forages play a vital role in protecting water supplies.

in a feedlot. As a result, much progress has been made in avoiding such problems. With forage crop production, the likelihood of such problems is particularly low.

When land has a thick cover of perennial forages, there is far less runoff, therefore less chance for fertilizers to be washed away. Pesticides are used relatively little in forage crop production. Most forage crops (especially perennial grasses) form dense root systems which effectively serve as filters to remove contaminants before they can seep into the ground water.

Waste Disposal

As environmental awareness increases, waste disposal becomes an increasingly important problem. Millions of tons of broiler litter are produced in the South each year. This and other animal waste requires disposal in environmentally acceptable ways.

A large percentage of the animal waste produced in the region is spread on forage crops. While beef and chicken may be in competition in the grocery store, they often make complementary enterprises on the farm, at least partially because poultry litter can be effectively used to provide nutrients for pasture growth. Evidence of this is the fact that in many areas of the South where broiler production has increased in recent years, so have the numbers of beef cattle.

Concern about safe disposal of other types of waste is on the increase as well. Disposal of municipal and industrial waste materials of various types is more and more frequently becoming a serious problem. Forage crops, which can absorb large quantities of applied effluent and remove nitrates and other potentially harmful materials from it over much of the year, are increasingly being used as a vegetative cover for waste disposal areas.

Air Quality

The "Greenhouse Effect" is often brought into discussions of ecology and the environment. In simple terms, this refers to more of the sun's energy reaching the earth and being held under a blanket of carbon dioxide and other pollutants rather than being deflected back into space. Substantial increases in the carbon dioxide concentration of the earth's atmosphere from burning of fossil fuels contribute to this effect.

Fortunately, during photosynthesis, plants consume carbon dioxide and release oxygen. While all types of plants accomplish this important function, the high leaf area of most forage canopies makes them particularly efficient.

Soil Improvement

The soil improvement characteristics of grasslands have long been recognized, but often seem to be forgotten or ignored on individual farms. After land has been devoted to perennial forages for several years, the trend is for subsequent annual crops to produce better than they would have if a sod-based rotation had not been used.

Even in areas which are not highly erodible, the use of forage crops in rotations can provide substantial benefits. Forages in rotation with row crops interrupt the cycle of weeds, diseases, insects, and nematodes while deep root penetration of some forage crops in compacted soils leaves root channels which enhance the root penetration of subsequent row crops. The crops which follow are productive, healthy, and are less likely to require expensive and potentially environmentally-harmful pesticide applications.

Perennial grasslands also tend to make the soil more suitable for subsequent crops in other ways. These include: improved tilth of the soil due to the activity of earthworms, soil insects, and microorganisms. Over time, the nutrient-holding capacity of the soil tends to increase and various mineral cycles operate to increase nutrient availability in the surface layer of soil.

Crop rotation is a basic desirable agronomic practice which tends to enhance long-term productivity of the soil. When forage legumes are used in rotations, following crops benefit from substantial quantities of residual N fixed by the legumes. Because of the numerous benefits from sod crops, many row crops should be grown in rotation with forage grasses or grass/legume mixtures.

Energy Inputs

Conservation of energy is also a concern of many environmentally-minded persons and, in general, the energy inputs into grassland systems are far less than for row crop production. Where perennials are grown, fuel-consuming tillage is generally limited to establishment. In other cases, as discussed in Chapter 12, no-tillage planting can be used to establish forage crops. Nutrient recycling from animals lowers the requirement for application of fertilizers as compared to other types of crop production, and pesticide usage tends to be lower on grasslands than other crop areas.

As indicated in **Figure 9.6,** animals sold off pasture remove a surprisingly small quantity of nutrients. Their bodies are primarily composed of carbon, hydrogen, and oxygen which come from the air and from water. Thus, conservation of energy is another benefit of grassland systems.

Enhancement Of Wildlife

Enhancement of wildlife populations with forage plants began in the South when Indians periodically burned patches of woodlands. The resulting development of open areas and stimulation of native grasses favored many types of wildlife and helped insure good hunting.

Many of the forage species normally planted for livestock in the South are also commonly planted for game animals (see Chapter 31). Deer and wild turkeys consume substantial quantities of green material, birds consume insects and seed of forage crops, and small animals such as rabbits use forage crops for food and cover.

Both game and nongame wildlife benefit from having access to forage crops. Wildlife populations are usually favored when a farm has a mixture of crops, pasture, and woodland. One of the joys of farm life is being able to observe animals and birds which survive and flourish along with agricultural enterprises.

Aesthetics and Appreciation

Many people find that there is something particularly restful, peaceful, and satisfying in watching animals graze. This is evidenced by the thousands of paintings and photographs which have pasture scenes as their primary focus. If the animals alone created this peaceful aura, a feed lot or corral scene would serve just as well.

Millions of people, many of whom may not have consciously

thought about it, appreciate a pastoral scene.

The key to developing a particularly keen appreciation of forage crops comes from learning more about them and from exposure to successful, positive examples. After all, this is consistent with the generally accepted idea that a person who has never been exposed to a particular type of music, art or food is unlikely to have developed much appreciation for it.

Certainly, forage crops beautify rural settings and are appreciated by many people. However, in addition to their visible beauty, knowledgeable persons can develop an appreciation for such things as: high yield; high forage quality; economy of production; low weed populations; little or no soil erosion; and a long productive season. Thus, a forage planting may be even more attractive to a person who fully realizes what it can do than to one who only sees its aesthetic beauty.

Appreciation of forage crop aesthetics and value is a seldom-mentioned aspect of forage/livestock enterprises, but it is likely more of a source of joy and satisfaction to many producers than they admit, or perhaps even realize. As knowledge of forages increases, so does this satisfaction. Producers who have an appreciation for beautiful, productive forages tend to grow beautiful, productive forages. In turn, they also tend to have efficient and profitable livestock operations.

Figure 32.3. What could be more noble and satisfying than spending a lifetime living on, and caring for, the land? A good farmer is a true conservationist.

Summary

Agricultural production is essential, but we also need to protect the environment. The number one environmental concern relating to agriculture is the protection of our soil and water resources. The growing of perennial pastures and forage crops is a particularly environmentally friendly type of agricultural production. In fact, it can be argued that agriculture and nature are rarely in better harmony than with grassland/animal production. ■

List of Appendices

A.1. Common Weights and Measures

Length

Inch = 1/12 or 0.083 foot = 2.54 centimeters = 25.4 millimeters
Foot = 12 inches = 0.3048 meters = 30.48 centimeters
Yard = 36 inches = 3 feet = 0.9144 meters
Rod = 16.5 feet = 5.5 yards = 5.03 meters
Furlong = 220 yards
Mile = 1,760 yards = 5,280 feet = 1.61 kilometers = 8 furlongs = 80 chains

Area

Square inch = 0.007 square foot = 6.45 square centimeters
Square foot = 144 square inches = 929.03 square centimeters
Square yard = 9 square feet = 0.836 square meters
Square rod = 30.25 square yards
Acre = 4,840 square yards = 43,560 square feet = 160 square rods = 4.047 square meters = 0.405 hectare
Hectare = 10,000 square meters = 2.47 acres
Square mile = 640 acres = 2.59 square kilometers = 1 section
Section = 1 square mile = 640 acres = 2.59 square kilometers

Liquid Measures

Teaspoon = 0.1667 fluid ounce = 80 drops = 4.93 milliliters
Tablespoon = 3 teaspoons = 0.5 fluid ounce = 14.8 milliliters
Fluid ounce = 2 tablespoons = 29.58 milliliters
Cup = 8 fluid ounces = 16 tablespoons = 236.6 milliliters
Pint = 2 cups = 16 fluid ounces = 473.2 milliliters
Quart = 4 cups = 2 pints = 32 fluid ounces = 0.946 liters
Liter = 2.113 pints = 1,000 milliliters = 1.057 quarts
Gallon = 4 quarts = 8 pints = 128 fluid ounces = 3.785 liters
Cubic foot of water = 7.5 gallons = 62.4 pounds = 28.3 liters
Acre inch of water = 27,154 gallons = 3,630 cubic feet.

Dry Measures

Teaspoon (level) = 0.35 cubic inch = 5.74 cubic centimeters
Tablespoon (level) = 1.05 cubic inch = 3 level teaspoons = 17.21 cubic centimeters
Cup = 16 level tablespoons = 16.8 cubic inches = 275.3 cubic centimeters
Pint = 2 cups = 32 level tablespoons = 33.6 cubic inches = 550.6 cubic centimeters
Quart = 2 pints = 64 tablespoons = 67.2 cubic inches = 1.101 liters
Peck = 8 quarts = 16 pints = 538 cubic inches = 8.8 liters
Bushel = 4 pecks = 2,150 cubic inches = 32 quarts = 35 liters

Volumes

Cubic inch = 0.00058 cubic foot = 16.4 cubic centimeters
Cubic foot = 1,728 cubic inches = 0.037 cubic yard = 0.028 cubic meter
Cubic yard = 27 cubic feet = 0.765 cubic meters

Weights

Gram = 15.43 grains = 1,000 milligrams
Ounce = 28.35 grams = 437.5 grains
Pound = 16 ounces = 7,000 grains = 454 grams
Kilogram = 1,000 grams = 2.205 pounds
Ton (short) = 2,000 pounds = 0.907 metric tons
Ton (long) = 2,240 pounds = 1.016 metric tons

A.2. Conversion Factors for English and Metric Units

To convert column 1 into column 2, multiply by	Column 1	Column 2	To convert column 2 into column 1, multiply by
	Length		
0.621	kilometer, km	mile, mi	1.609
1.094	meter, m	yard, yd	0.914
0.394	centimeter, cm	inch, in	2.54
	Area		
0.386	kilometer^2, km^2	mile^2, mi^2	2.590
247.1	kilometer^2, km^2	acre, acre	0.00405
2.471	hectare, ha	acre, acre	0.405
	Volume		
0.00973	cubic meter, m^3	acre-inch	102.8
3.532	hectoliter, hl	cubic foot, ft^3	0.2832
2.838	hectoliter, hl	bushel, bu	0.352
0.0284	liter, l	bushel, bu	35.24
1.057	liter, l	quart (liquid), qt	0.946
	Mass		
1.102	ton (metric)	ton (English)	0.9072
2.205	quintal, q	hundredweight, cwt (short)	0.454
2.205	kilogram, kg	pound, lb	0.454
0.035	gram, g	ounce (avdp), oz	28.35
	Pressure		
14.50	bar	lb/inch^2, psi	0.06895
0.9869	bar	atmosphere, atm	1.013
0.9678	kg (weight)/cm^2	atmosphere, atm	1.013
14.22	kg (weight)/cm^2	lb/inch^2, psi	0.07031
14.70	atmosphere, atm	lb/inch^2, psi	0.06805
	Yield or Rate		
0.446	ton (metric)/hectare	ton (English)/acre	2.240
0.891	kg/ha	lb/acre	1.12
0.891	quintal/hectare	hundredweight/acre	1.12
1.15	hectoliter/hectare, hl/ha	bu/acre	0.87
	Temperature		
$(1.8 \times C) + 32$	Celsius, C	Fahrenheit, F	0.56 (F-32)
	-17.8°	0°F	
	0°C	32°F	
	20°C	68°F	
	100°C	212°F	

Metric Prefix Definitions

mega	1,000,000	deca	10	centi	0.01
kilo	1,000	basic metric unit	1	milli	0.001
hecto	100	deci	0.1	micro	0.000001

A.3. Helpful Calculations and Formulas

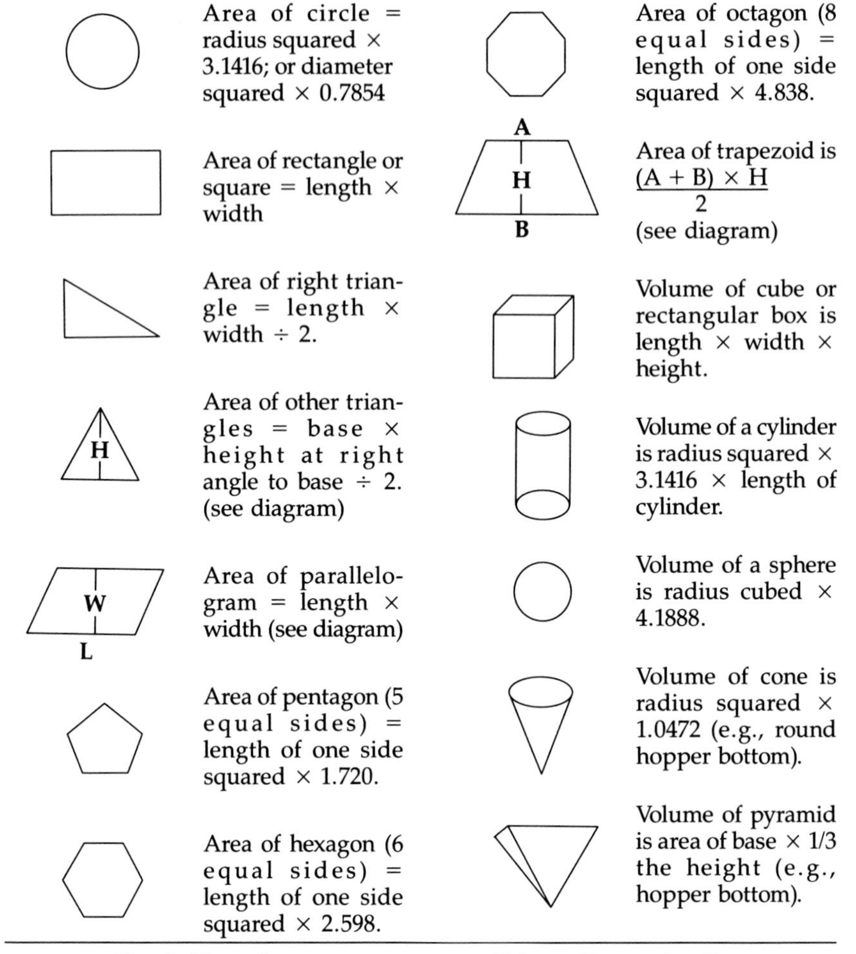

Area of circle = radius squared × 3.1416; or diameter squared × 0.7854

Area of rectangle or square = length × width

Area of right triangle = length × width ÷ 2.

Area of other triangles = base × height at right angle to base ÷ 2. (see diagram)

Area of parallelogram = length × width (see diagram)

Area of pentagon (5 equal sides) = length of one side squared × 1.720.

Area of hexagon (6 equal sides) = length of one side squared × 2.598.

Area of octagon (8 equal sides) = length of one side squared × 4.838.

Area of trapezoid is $\frac{(A + B) \times H}{2}$ (see diagram)

Volume of cube or rectangular box is length × width × height.

Volume of a cylinder is radius squared × 3.1416 × length of cylinder.

Volume of a sphere is radius cubed × 4.1888.

Volume of cone is radius squared × 1.0472 (e.g., round hopper bottom).

Volume of pyramid is area of base × 1/3 the height (e.g., hopper bottom).

Handy Formulas

To find circumference of a circle when diameter is known, multiply diameter by 3.1416.

To find diameter when circumference is known, divide circumference by 3.1416 or multiply by 0.3183.

Volume Conversion Factors

Cu. ft. × 0.8 equals bushels grain or shelled corn
Cu. ft. × 0.4 equals bushels ear corn
Cu. ft. × 7.48 equals gallons
Cu. ft. × 62.4 equals pounds water
Gallons × 8.330 equals pounds water
Gallons × 0.1337 equals cu. ft.
Cu. in. ÷ 1,728 = cu. ft.
Cu. yd. × 27 = cu. ft.
Cu. ft. ÷ 27 = cu. yds.

A.4. Calculating Fertilizer and Seed Weights for Small Areas

lb/A	grams/sq. ft.	lb/sq. ft.	lb/A	grams/sq. ft.	lb/sq. ft.
1	.01042	.000023	75	.7815	.00172
2	.02064	.000046	100	1.042	.0023
3	.03126	.000069	150	1.563	.0035
4	.04168	.000091	200	2.084	.0046
5	.05210	.000114	300	3.126	.0069
6	.06252	.000138	400	4.168	.0092
7	.07294	.000161	500	5.210	.0115
8	.08336	.000184	600	6.252	.0138
9	.09378	.000207	700	7.294	.0141
10	.1042	.00023	800	8.336	.0184
15	.1565	.00034	900	9.378	.0207
25	.2605	.00057	1,000	10.42	.0230
50	.5210	.00115	1 sq. ft. = .00002296 acre		

A.5. Estimating Acres within a Field[1]

W = 300 feet

L = 1,742 feet

$$\text{Acres} = \frac{\text{length (L)} \times \text{width (W)}}{43{,}560}$$

Example:

$$\frac{1{,}742 \times 300}{43{,}560} = 11.997 = 12 \text{ acres}$$

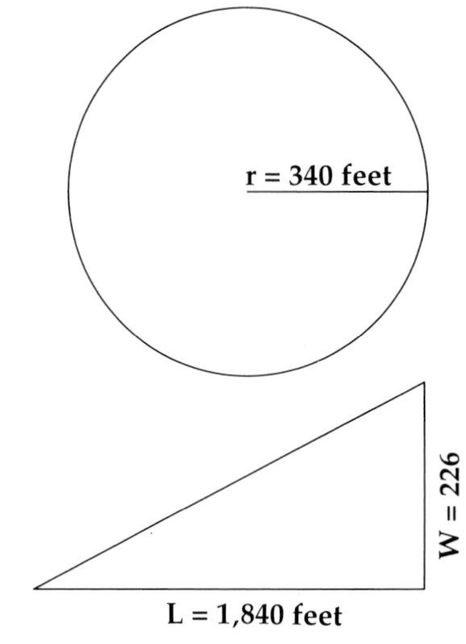

$$\text{Acres} = \pi\ r^2/43{,}560$$

π = 3.14

r = radius

$$\text{Example: } \frac{3.14 \times 340^2}{43{,}560} = 8.33 \text{ acres}$$

$$\text{Acres} = \frac{\tfrac{1}{2}\,(\text{length} \times \text{width})}{43{,}560}$$

$$\text{Example: } \frac{\tfrac{1}{2}\,(1{,}840 \times 226)}{43{,}560} = 4.77 \text{ acres}$$

[1]Most fields will not exactly fit the above description and may require a combination of the above or use of other formulas.

A.6. Calibrating Seeders

To calibrate any seeding machine, two things must be known: 1) area covered, and 2) amount of seed used.

$$\text{Seeding rate (SR)} = \frac{\text{S (amount of seed)}}{\text{A (area covered in acres)}}$$

$$A = \frac{\text{(area covered in square feet)}}{\text{43,560 (square feet in one acre)}}$$

1 lb = 16 oz = 454 grams

1. Area covered.
   A. Drill to determine area covered
      1. measure seeding width of machine used
      2. a) measure distance around drive wheel or sprocket and turn a given number of times
         b) pull the machine at the desired seeding speed for some measured distance

2. Amount of seed used.
   A. Place some type of collection device or container at seed outlet(s) collect and weigh seed

Examples:

1. A drill with a 7 foot wide seeding width is pulled 100 feet with 0.25 lb of seed dispensed. What is the seeding rate?

$$A = \frac{7 \times 100}{43{,}560} = 0.016$$

$$SR = \frac{S = 0.25}{A = 0.016} = 15.6 \text{ lb/A}$$

2. A cyclone seeder with a 12 foot seeding width dispenses 0.12 lb of seed in 100 feet. Calculate the seeding rate.

$$SR = S/A$$

$$A = \frac{12 \times 100}{43{,}560} = 0.028$$

$$SR = \frac{S = 0.12}{A = 0.028} = 4.3 \text{ lb/A}$$

3. Three lb of seed were placed in a spinner (cyclone) seeder with a 12 foot seeding width. Seed was sufficient to cover a distance of 1,000 feet. What is the seeding rate per acre?

$$SR = S/A$$

$$A = \frac{12 \times 1{,}000}{43{,}560} = 0.28$$

$$SR = \frac{3}{0.28} = 10.7 \text{ lb/A}$$

A.7. Composition of Principal Fertilizer Materials

Material Supplying	Nitrogen (N) %	Phosphate (P_2O_5) %	Potash (K_2O) %	Sulfur (S) %	Avail- ability[1]	Equivalent Acidity (A) or Basicity (B) lb $CaCO_3$/100 lb Product[2]
Nitrogen						
Ammonium nitrate	33 to 34	0	0	0	Q	119 to 122AA
Ammonium nitrate and limestone	20.5	0	0	0	Q	Neutral
Anhydrous ammonia	82	0	0	0	Q	295A
Calcium nitrate	15.5	0	0	0	Q	20B
Sodium nitrate	16	0	0	0	Q	29B
Urea-ammonium nitrate solution	28 to 32	0	0	0	Q	101 to 115A
Ammonium sulfate	21	0	0	24	Q	151A
Urea	46	0	0	0	Q	166A
Ammonium thiosulfate	12	0	0	26	Q	80a
Sewage sludge (activated)	4 to 6	2.5 to 4	0	<1	M	18A
Phosphorus						
Ammonium polyphosphate (APP)	10	34	0	0	Q	73A
Diammonium phosphate (DAP)	18	46	0	0	Q	102A
Monammonium phosphate (MAP)	10 to 12	50 to 55	0	0	Q	90 to 99A
Normal superphosphate	0	18 to 20	0	12	Q	Neutral
Triple Superphosphate	0	44 to 46	0	1	Q	Neutral
Ground rock phosphate	0	26 to 35 (Approx. 3% available)	0	0	S	10B
Basic slag	0	10 to 25	0	0	Q	70 to 90B
Potassium						
Potassium chloride	0	0	60 to 62	0	Q	Neutral
Potassium nitrate	13	0	44	0	Q	26B
Potassium sulfate	0	0	48 to 52	18	Q	Neutral
Sulfate of potash magnesia	0	0	22	22	Q	Neutral

[1]Availability ratings: Quick (Q); Medium (M); Slow (S).

[2]Fertilizers have either neutral, acidic (lower soil pH), or basic (increase soil pH) effects when added to the soil. These effects are commonly expressed in terms of the amount of pure calcium carbonate that would be required to either offset the acid-forming reactions of 100 lb of the fertilizer material or the amount of calcium carbonate required to equal the acid-neutralizing effects of 100 lb of the fertilizer. Most of the acid-forming effects are due to the activities of soil bacteria which convert ammonium-nitrogen to nitrite and nitrate in the process of nitrification. The values in this table are based on the theoretical values for acidity produced during nitrification.

Source: Potash & Phosphate Institute; adapted from various sources.

A.8. Forage Quality Parameters for Selected Forage Crops[1]

Crop	CP[2]	ADF	NDF	TDN	RFV
Alfalfa					
Bud	22-26	28-32	38-47	64-67	127-164
Early flower	18-22	32-36	42-50	61-64	113-142
Mid bloom	14-18	36-40	46-55	58-61	98-123
Corn Silage					
Well eared	7-9	23-30	48-58	66-71	105-138
Mid to poorly eared	7-9	30-39	58-67	59-66	81-105
Cool Season Perennial Grasses (Tall Fescue/Orchardgrass)					
Vegetative-boot	12-16	30-36	50-56	61-66	101-122
Boot-head	8-12	36-42	56-62	56-61	84-101
Ryegrass					
Vegetative-boot	12-16	27-33	47-53	63-68	111-134
Boot-head	8-12	33-39	53-59	59-63	92-111
Switchgrass/caucasian bluestem					
< Boot	10-14	35-40	55-60	58-62	90-104
Mature, head	6-10	40-50	60-75	50-58	62-90
Bermudagrass					
4 week old	10-12	33-38	63-68	58-62	81-93
8 week old	6-8	40-45	70-75	45-50	67-77
Pearl millet, sorghum-sudangrass	8-14	35-40	55-70	58-62	77-104
Red Clover					
Early flower	14-16	28-32	38-42	64-67	142-164
Late flower	12-14	32-38	42-50	59-64	110-142
Annual Lespedeza	12-16	35-40	45-55	58-62	98-127

[1]These are estimates. Forage quality varies as a result of many factors (see Chapter 16). Source: Dr. Jimmy Henning and Dr. Garry Lacefield, Extension Forage Specialists, University of Kentucky.

[2]Abbreviations over columns are as follows: CP = crude protein; ADF = acid detergent fiber; NDF = neutral detergent fiber; TDN = total digestible nutrients; RFV = relative feed value.

A.9. Yield Estimates for Three Pasture Types of Various Heights

	Pasture canopy height (inches)			
	2	4	6	8
Species	DM yield, lb/A			
Fescue-clover	700	1,500	2,000	2,400
Hybrid bermuda	1,000	2,000	2,500	3,000
Rye (small grain)	300	700	1,300	1,700

Source: Chamblee, Douglas S. and James T. Green, Jr., eds., 1995 Production and Utilization of Pastures and Forages in North Carolina. North Carolina Ag. Res. Ser. Tech. Bull. 305.

A.10. Score Card for Hay Quality Evaluation

			Possible Score	Your Score
I.	**Stage of Harvest**	1. Before blossom or heading	26-30	____
		2. Early blossom or early heading	21-25	____
		3. Mid-to-late bloom or head	16-20	____
		4. Seed stage (stemmy)	11-15	____
II.	**Leafiness**	1. Very leafy	26-30	____
		2. Leafy	21-25	____
		3. Slightly stemmy	16-20	____
		4. Stemmy	11-15	____
III.	**Color**	1. Natural green color of crop	13-15	____
		2. Light green	10-12	____
		3. Yellow to slightly brownish	7-9	____
		4. Brown or Black	0-6	____
IV.	**Odor**	1. Clean–"crop odor"	13-15	____
		2. Dusty	10-12	____
		3. Moldy–mousey or musty	7-9	____
		4. Burnt	0-6	____
V.	**Softness**	1. Very soft and pliable	9-10	____
		2. Soft	7-8	____
		3. Slightly harsh	5-6	____
		4. Harsh, brittle	0-4	____
		Sub-total		____
VI.	**Penalties**	1. Trash, weeds, dirt, and other foreign material	0-35 minus	
			Total	____

Scoring

90 and above	–Excellent hay
80-89	–Good hay
65-79	–Fair hay
Below 65	–Poor hay

Source: J.D. Burns and G.D. Lacefield 1991. Southeast Regional Beef Cow-Calf Handbook SR-5004, 1991.

A.11. Score Card for Hay Crop Silage Evaluation

			Possible Score	Your Score
I.	**Stage of Harvest**	1. Before blossom or heading	36-40	____
		2. Early blossom or early heading	28-35	____
		3. Mid-to-late bloom or head	16-27	____
		4. Seed stage	0-15	____
II.	**Color**	1. **Desirable**–Light to dark green depending on crop and/or additive used. Legumes may have a darker color.	9-12	____
		2. **Acceptable**–Yellowish-green to brown.	5-8	____
		3. **Undesirable**–Dark brown or black indicating excessive heating or putrefaction. Predominantly white or gray, indicating excessive mold development.	0-4	____
III.	**Odor**	1. **Desirable**–Pleasant with no indication of putrefaction.	24-28	____
		2. **Acceptable**–Slight burnt odor, fruity, yeasty, or musty, which indicates slightly improper fermentation.	11-23	____
		3. **Undesirable**–Strong burnt odor indicates excessive heating; a putrid odor with sliminess indicates improper fermentation.	0-10	____
IV.	**Moisture**	1. No free water when squeezed in hand. Well preserved silage.	18-20	____
		2. Some moisture can be squeezed from silage or silage somewhat dry and musty.	9-17	____
		3. Silage wet, slimy or soggy, water easily squeezed from sample. Silage too dry with a strong burnt odor.	0-10	____
			Total	____

Scoring

90 and above	–Excellent hay
80-89	–Good hay
65-79	–Fair hay
Below 65	–Poor hay

Source: Adapted from J.D. Burns, Extension Agron. (ret.), University of Tennessee, April, 1991.

A.12. Score Card for Corn Silage Evaluation

			Possible Score	Your Score
I.	**Grain Content**	1. High–35% and above	36-40	____
		2. Medium–15 to 35%	28-35	____
		3. Low–1 to 14%	16-27	____
		4. None (either no ears developed or ears removed).	0-15	____
II.	**Color**	1. **Desirable**–green to yellowish-green.	9-12	____
		2. **Acceptable**–Yellow to brownish.	5-8	____
		3. **Undesirable**–Deep brown or black indicating excessive heating or putrefaction. Predominantly white or gray–excessive mold development.	0-4	____
III.	**Odor**	1. **Desirable**–Light, pleasant odor; no indication of putrefaction.	24-28	____
		2. **Acceptable**–Fruity, yeasty, or musty, which indicates slightly improper fermentation. Slight burnt odor. Sharp vinegar odor.	11-23	____
		3. **Undesirable**–Strong burnt odor indicating excessive heating. Putrid, indicating improper fermentation. A very musty odor, indicating excessive mold which is readily visible throughout silage.	0-10	____
IV.	**Moisture**	1. No free water when squeezed in hand. Well preserved silage.	9-10	____
		2. Some moisture can be squeezed from silage or silage dry and musty.	5-8	____
		3. Silage wet, slimy or soggy, water easily squeezed from sample. Silage too dry with a strong burnt odor.	0-4	____
V.	**Chop**	1. Small, uniform, sharp angled pieces of silage.	9-10	____
		2. Silage uniform in cut, but slightly stringy, some large pieces of shucks, cobs, stalks.	5-8	____
		3. Silage stringy, puffy or large variable sized pieces.	0-4	____

Scoring 90 and above–Excellent silage; 80-90–Good silage; 65-79–Fair silage; Below 65–Poor silage. **Total** ____

Source: Adapted from J.D. Burns, Extension Agron. (ret.), University of Tennessee, Nov., 1990.

A.13. Determining Forage Moisture Content Using a Microwave Oven

1. Chop fresh forage into 1 to 2 inch lengths for ease of handling.
2. Weigh out approximately 100 grams (3.5 ounces) of chopped forage.
3. Spread forage thinly on a microwave-safe dish and place into microwave.
4. Heat for 2 minutes and reweigh.
   a) If forage is not completely dry, reheat for 30 seconds and reweigh. (Microwaves vary considerably in drying capacity. It is better to dry for short intervals and reweigh until the last two weights are constant, than to overdry and run the risk of burning and damage to oven.) Continue this process until back-to-back weights are the same or charring occurs.
   b) If charring occurs, use the previous weight.
5. Calculate moisture content using the following equation:

$$\% \text{ Moisture Content} = \frac{W1 - W2 \times 100}{W1}$$

W1 = weight of forage before heating
W2 = weight of forage after heating

Dry matter (DM) is the percentage of forage that is not water. DM equals 100% minus percent water.

Example: moisture content 14%

DM = 100 − 14 = 86%

Results on an "as-fed basis" reflect total nutrient concentration including water of sample analyzed or to be fed.

A.14. Wind-Chill Factor

A strong wind makes a cold day colder. This effect is called the wind-chill factor.

The chart below gives examples of wind-chill equivalents, as measured by the National Weather Service. To determine the wind-chill temperature, find the outside air temperature on the top line, then read down the column to the measured wind speed. For example, when the outside air temperature is 30°F and the wind speed is 15 miles per hour, the wind-chill temperature is 9°F.

Wind Speed (miles per hour)	Equivalent Temperature (°F)								
Calm	35	30	25	20	15	10	5	0	−5
5	32	27	22	16	11	6	0	−5	−10
10	22	16	10	3	−3	−9	−15	−22	−27
15	16	9	2	−5	−11	−18	−25	−31	−38
20	12	4	−3	−10	−17	−24	−31	−39	−46
25	8	1	−7	−15	−22	−29	−36	−44	−51
30	6	−2	−10	−18	−25	−33	−41	−49	−56
35	4	−4	−12	−20	−27	−35	−43	−52	−58

A.15. Nitrates in Forages

Nitrate in feed on dry matter basis %	ppm		Comments
0.0-0.25	0-2,500	SAFE	Generally considered **SAFE**.
0.25-0.50	2,500-5,000	CAUTION	Generally safe when fed with a balanced ration. For pregnant animals limit to one-half of total dry ration. Make certain water for livestock is low in nitrates. Prolonged feeding may result in Vitamin A deficiency. Do not feed with liquid feed or other non-protein nitrogen supplements. Be cautious with pregnant and young animals.
0.50-1.5	5,000-15,000	DANGER	Limit to one-fourth of ration. Should be well fortified with energy, minerals, and Vitamin A. May experience milk production loss in 4 to 5 days, possible occurrence of reproduction problems.
Over 1.5	over 15,000	TOXIC	Toxic. Do not use in free-choice feeding program. Feed containing such levels may be ground and mixed if high nitrate feed is limited to 15% of total ration.

Method of expression	Chemical designation	To convert to nitrate multiply by	To convert to nitrate nitrogen multiply by	To convert to potassium nitrate multiply by
Nitrate	NO_3	1.00	0.23	1.63
Nitrate nitrogen	NO_3-N	4.40	1.00	7.20
Potassium nitrate	KNO_3	0.61	0.14	1.00

Example: 1.0% nitrate × 0.23 = 0.23% nitrate nitrogen
1.0% potassium nitrate × 0.14 = 0.14% nitrate-nitrogen
To convert ppm to percent, move the decimal point four places to the left, e.g. 4,400 ppm = 0.44%,
5,000 ppm = 0.5%.

NOTE: The table above presents general guidelines which can be helpful in determining the nitrate toxicity potential of forages. However, the likelihood of toxicity can be affected by several factors including species and class of animal, degree of animal stress, animal health, extent of previous exposure to nitrates, and feeding management (see Chapter 21). In addition, sampling errors can result in misleading results. Considerable variation exists among state-to-state guidelines with regard to nitrate levels which may cause animal feeding problems.

Source: Materials compiled from several sources including: D. Miksch and G.D. Lacefield. 1988. Kentucky Agric. Ext. Ser. Herd Health Memo and B.G. Ruffin, D.M. Ball, and H.A. Kjar. 1979. Alabama Agric. Ext. Ser. Cir. ANR 112.

A.16. Silo Capacities–Upright

Tower silo capacity, wet tons

Silage at 65% moisture, wet basis. Capacities allow 1 ft. of unused depth for settling in silos up to 30 ft. high and 1 ft. more for each 10 ft. beyond 30 ft. height. Capacities rounded to nearest 5 tons.

Silo height ft.	Silo diameter, ft.								
	14	16	18	20	22	24	26	28	30
	wet tons								
20	45	60	75	95	115	135	160	185	210
24	60	75	95	125	150	175	205	235	275
28	75	100	125	150	185	215	255	295	340
32	90	115	150	185	225	265	310	365	415
36	105	135	175	215	265	310	370	430	490
40	125	165	205	255	305	365	430	495	570
44	145	185	235	290	350	420	490	570	655
48	160	210	265	330	400	475	560	645	745
52	185	235	300	370	450	530	625	725	830
56	205	265	335	410	500	590	695	805	925
60	225	290	370	455	550	650	780	885	1020
64			405	500	600	715	850	970	1120
68			445	545	650	780	925	1060	1215
72						840	1000	1145	1310
76						890	1075	1220	1400
80						955	1120	1300	1485
84							1215	1385	1575
88							1285	1475	1660
92							1375	1560	1745
96							1445	1640	1830
100							1515	1730	1930

Tower silo capacity, tons dry matter

Capacities allow 1 ft. of unused depth for settling in silos up to 30 ft. high and 1 ft. more for each 10 ft. beyond 30 ft. height. To determine silo capacity in "wet tons", multiply silo capacity value, dry matter, by the DMF value from Appendix A.18.B. Capacities rounded to nearest 5 tons.

Silo height ft.	Silo diameter, ft.								
	14	16	18	20	22	24	26	28	30
	wet tons								
20	15	20	25	35	40	45	55	65	75
24	20	25	35	45	50	60	70	85	95
28	25	35	45	55	65	75	90	105	120
32	30	40	50	65	80	95	110	125	145
36	35	50	60	75	90	110	130	150	170
40	45	55	70	90	105	125	150	175	200
44	50	65	80	100	125	145	170	200	230
48	55	75	95	115	140	165	195	225	260
52	65	85	105	130	155	185	220	255	290
56	70	95	115	145	175	205	245	280	325
60	80	100	130	160	190	230	275	310	355
64			140	175	210	250	300	340	390
68			155	190	230	270	325	370	425
72						295	350	400	460
76						315	375	425	490
80						335	400	455	520
84							425	485	550
88							450	515	580
92							480	545	610
96							505	575	640
100							530	605	675

Source: Reproduced with permission from: Beef Housing and Equipment Handbook, MWPS-6, 4th Edition, 1987. Midwest Plan Service, Ames, Iowa.

A.17. Silo Capacities–Horizontal

Horizontal silo capacity, wet tons

65% moisture; 40 lb/ft.^{3} or 50 ft.^{3} = 1 ton; 1.25 ft.^{3}/bu. Silo assumed level full. Capacities rounded to nearest 5 tons. To calculate capacity of other silo sizes: (silage depth, ft. × silo width, ft. × silo length, ft.) ÷ 50.

Depth ft.	Silo floor width, ft. 20	30	40	50	60	70	80	90	100
	wet tons/10 ft. length								
10	40	60	80	100	120	140	160	180	200
12	50	70	95	120	145	170	190	215	240
14	55	85	110	140	170	195	225	250	280
16	65	95	130	160	190	225	255	290	320
18	70	110	145	180	215	250	290	325	360
20	80	120	160	200	240	280	320	360	400

Horizontal silo capacity, dry matter

Silo assumed level full. Capacities rounded to nearest 5 tons.

Depth ft.	Silo floor width, ft. 20	30	40	50	60	70	80	90	100
	tons dry matter/10 ft. length								
10	15	20	30	35	40	50	55	65	70
12	15	25	35	40	50	60	65	75	85
14	20	30	40	50	60	70	80	90	100
16	20	35	45	55	65	80	90	100	110
18	25	40	50	65	75	90	100	115	125
20	30	40	55	70	85	100	110	125	140

Source: Reproduced with permission from: Beef Housing and Equipment Handbook, MWPS-6, 4th Edition, 1987. Midwest Plan Service, Ames, Iowa.

A.18. Silage Conversions

A. Grain moisture factor (GMF)

To convert wet tons to tons at 15.5% moisture, divide by GMF. To convert tons at 15.5% moisture to wet tons, multiply by GMF. GMF = 84.5 ÷ (100 − % moisture).

% moisture	MCF
18	1.03
20	1.06
22	1.08
24	1.11
26	1.14
28	1.17
30	1.21
32	1.24
34	1.28
36	1.32
38	1.36
40	1.41
45	1.54
50	1.69

B. Dry matter factor (DMF)

To convert from wet tons to tons of dry matter, divide by the DMF. To convert from tons of dry matter to wet tons, multiply by the DMF. DMF = 100 ÷ (100 − % moisture).

% moisture	DMF
30	1.43
40	1.67
50	2.00
55	2.22
60	2.50
65	2.86
70	3.33
75	4.00
80	5.00

Source: Reproduced with permission from: Beef Housing and Equipment Handbook, MWPS-6, 4th Edition, 1987. Midwest Plan Service, Ames, Iowa.

A.19. Estimating Hay Needs

Type livestock	Approximate lb hay/animal/day
Dry, pregnant cows	15-20
Cows with calves	25-28
Replacement heifers	10-12
Bred yearling heifers	18-23
Herd bull	28-30
Stocker steers	10-14
Horse	24-30
Sheep	3-6
Goat	2-5
Fallow deer	2-5

Total lb/Day × Days of Winterfeeding = Total Pounds of Hay Needed
Total lb Needed ÷ 2,000 = Tons Needed

A.20. Gestation Periods

Species	Days[1]
Cat	63
Cattle	282
Coyote	63
Deer	210
Dog	63
Donkey	365
Fox	52
Goat	148
Guinea Pig	68
Horse	335
Llama	342
Pig	114
Rabbit	32
Rat	22
Sheep	148
Squirrel	44

[1]These are approximate values based on information from various sources. Gestation periods may vary within species due to breed and/or environmental conditions.

A.21. Incubation Periods

Species	Days[1]
Chicken	20-22
Duck	26-28
Goose	26-28
Pheasant	21-28
Pigeon	16-18
Quail	21-28
Turkey	26-28

[1]These are approximate values based on information from several sources. Incubation periods can vary within species due to breed and/or environmental conditions.

A.22. Names of Breeding Animals and Their Offspring

Species	Bovine	Ovine	Caprine	Equine	Porcine	Feline	Canine
Common name	**Cattle**	**Sheep**	**Goat**	**Horse**	**Pig**	**Cat**	**Dog**
Male	Bull	Ram	Buck	Stallion	Boar	Tom	Dog
Female	Cow	Ewe	Doe	Mare	Sow	Queen	Bitch
Name of young	**Calf**	**Lamb**	**Kid**	**Foal**	**Pig**[1]	**Kitten**[1]	**Puppy**[1]
Young male	Bull	Ram lamb	Buckling	Colt	Boar	—	—
Young female	Heifer	Ewe lamb	Doeling	Filly	Gilt	—	—
Castrated male	Steer	Wether	Wether	Gelding	Barrow	—	—

[1]These are born in litters.

A.23. Characteristics of Forage Grasses

Name	Climate zone adaptation[1]	Seedling vigor	Tolerance[2] to			
			Soil acidity	Poor drainage	Drought	Grazing
Warm season perennial grasses						
Bahiagrass	AB	P	E	G	E	E
Bermudagrass	ABC	F	E	P	E	E
Dallisgrass	AB	P	F	E	G	G
Johnsongrass	ABCD	G	F	E	G	P
Switchgrass	BCD	P	F	F	E	P
Warm season annual grasses						
Corn	ABCD	E	F	P	P	P
Pearl millet	ABCD	E	E	P	E	F
Sorghum	ABCD	G	P	P	E	F
Sorghum-sudan	ABCD	E	P	F	G	F
Cool season perennial grasses						
Kentucky bluegrass	CD	P	F	F	P	E
Orchardgrass	BCD	F/G	F	F	F	G
Reed canarygrass	CD	F	G	E	G	F
Tall fescue E+[3]	BCD	G	G	G	G	E
Tall fescue E−[3]	BCD	F/G	G	G	F	G
Timothy	CD	G	F	F	F	F
Cool season annual grasses						
Annual ryegrass	ABC	G	G	E	F	E
Oats	ABC	E	F	F	F	G
Rye	ABCD	E	E	F	F	G
Wheat	ABCD	E	P	P	F	G

[1]Some species may be adapted only in portions of a zone.
[2]E=Excellent, G=Good, F=Fair, P=Poor
[3]E+=endophyte-infected, E−=endophyte-free

A.24. Characteristics of Forage Legumes

Name	Climate zone adaptation[1]	Seedling vigor	Tolerance[2] to Soil acidity	Poor drainage	Drought	Grazing
Warm season perennial legumes						
Perennial peanut	A	[3]	G	P	G	F
Sericea lespedeza	ABCD	P	E	F	E	P
Warm season annual grasses						
Annual lespedeza	BCD	F	E	F	G	G
Cool season perennial legumes						
Alfalfa	ABCD	G	P	P	E	P[4]
Birdsfoot trefoil	BCD	P	G	G	G	F
Red clover	ABCD	E	F	F	F/G	G
White clover	ABCD	F	F	G	P	E
Cool season annual legumes						
Arrowleaf clover	AB	F	F	P	F	G
Berseem clover	A	G	P	G	G	F
Caley pea	AB	G	F	G	F	F
Crimson clover	ABC	G	G	P	F	F
Hairy vetch	ABCD	E	G	P	F	F
Rose clover	A	F	P	P	G	F
Subterranean clover	A	G	F	G	F	E

[1]Some species adapted only in portions of a zone.
[2]E=Excellent, G=Good, F=Fair, P=Poor
[3]Vegetatively propagated
[4]Grazing-tolerant varieties are rated G

A.25. Water Requirements for Various Animal Groups

	Daily needs, gal/head[1] 50°F	90°F
Beef		
400 lb calf	4	10
800 lb feeder	7	15
1,000 lb feeder	8	17
Cows and bulls	8	20
Dairy		
Cows	15	30
Calves	2	12
Replacement heifers	64	15
Bulls	8	20
Horses or Mules	8	12
Swine (per 100 lb liveweight)	1	1.5
Turkeys (per 100 head)	10	15
Chickens (per 100 layers)	6	9
Fallow Deer	1	1.5

[1]Water consumption varies considerably depending on animal class, animal overall health, temperature, stage of lactation and other environmental conditions. A rule of thumb of 1 gal. of water per 100 lb of body weight is used by many cattlemen. One gal H_2O = 8.34 lb; one cubic foot of H_2O = 62.4 lb. Another formula used is: Water consumption = 0.39 (temp) − 8.87.

A.26. Stages of Maturity

	GRASSES
Vegetative	leafy growth, few stems, no reproductive (seedhead) growth
Late vegetative	stem elongates
Boot	stem elongated, top of stem swollen
Early bloom	seed heads (flower heads) begin to emerge
Mid-bloom	at least 25% of seed emerged, pollen beginning to shed
Full bloom	most seedheads emerge; peak pollen shed
Milk	all seedheads emerge; seed forming, seed soft and immature
Dough	seed becoming harder and have a doughlike consistency
Mature	seed ready for harvest
	LEGUMES
Vegetative	leaf and stem growth; no buds, flowers, or seed pods
Bud	buds begin to swell and become apparent at a few nodes
Late bud	several nodes with buds; buds more swollen
Early flower	a few buds open, flower color apparent
Late flower	many flowers apparent
Early seed	green seed pods apparent on a few flowers
Late seed	many green seed pods apparent; some seed pods turning brown
Mature	seed pods brown to black and dry; ready to harvest as moisture content permits

A.27. Essential Elements for Plants and Animals

		Plant	Animal
Carbon	C	✔	✔
Hydrogen	H	✔	✔
Oxygen	O	✔	✔
Nitrogen	N	✔	✔
Phosphorus	P	✔	✔
Potassium	K	✔	✔
Calcium	Ca	✔	✔
Magnesium	Mg	✔	✔
Sulfur	S	✔	✔

		Plant	Animal
Arsenic	As	–	✔
Boron	B	✔	?
Chloride	Cl	✔	✔
Chromium	Cr	–	?
Cobalt	Co	✔	✔
Copper	Cu	✔	✔
Fluorine	F	–	✔
Iodine	I	–	✔
Iron	Fe	✔	✔
Manganese	Mn	✔	✔
Molybdenum	Mo	✔	✔
Nickel	Ni	✔	–
Selenium	Se	–	✔
Silicon	Si	✔	✔
Sodium	Na	✔	✔
Vanadium	Va	–	✔
Zinc	Zn	✔	✔

A.28. Grazing Equations

$$\textbf{Number of Paddocks} = \frac{\textbf{Days of rest}}{\textbf{Days of grazing}} + 1$$

Example: Graze 4 days, rest 28 days

$$\frac{28}{4} + 1 = 8$$

$$\textbf{Acres Required per Paddock} = \frac{\text{Avg. wt. of animals to be grazed} \times \text{Dry matter consumed}^{1}\text{ per animal as \% of body weight} \times \text{Number of animals} \times \text{Days on the pasture}}{\text{Dry matter available in the area to be grazed}^{2} \times \text{\% of the dry matter utilized by grazing}^{3}}$$

Example: Thirty 600 lb steers consuming 3% body weight for 4 days. Twelve inches growth, thick stand (12 × 225 lb/inch) utilizing approximately 60%.

$$\frac{600 \times .03 \times 40 \times 4}{(12 \times 225) \times .60} = \frac{2{,}880}{1{,}620} = 1.8$$

Total Acres Required = Number of Paddocks × Acres required per paddock

Example: 8 × 1.8 = 14.4

$$\textbf{Stocking Rate} = \frac{\textbf{No. animals to be grazed}}{\textbf{Total acres grazed}}$$

Example: $\frac{40}{14.4} = 2.8$

$$\textbf{Stocking Density} = \frac{\textbf{No. animals grazing}}{\textbf{Paddock size (acres)}}$$

Example: $\frac{40}{1.8} = 22$

---

[1]Grazing animals will normally consume from 2 to over 3 percent of their body weight in dry matter each day. Intake varies with pasture quality and quantity and animal class. Dry beef cows will normally consume around 2 percent per day, growing steers and lactating dairy cows will normally consume around 3 percent.

[2]Dry matter available depends on species, growing conditions and density. Although considerable variation will occur, a range of 150 to 300 pounds of dry matter for each inch of usable pasture growth is often used.

[3]Amount of dry matter utilized will depend on grazing method and grazing time. A range of 40 to 70 percent pasture utilization is common. Utilization in the upper portion of this range may occur with higher stocking rates and shorter grazing times. By comparison, forage utilization by other means might be assumed to be as follows: hay, 70 to 80 percent; strip grazing, 75 to 80 percent; silage or greenchop, 85 to 95 percent (see **Appendix A.9**).

A.29. Compatibility[1] of Legumes with Grasses

Legume	Bahiagrass or bermudagrass	Dallisgrass	Johnsongrass	Tall fescue, orchardgrass, timothy, or Ky. bluegrass	Small grain or ryegrass
Perennial peanut	X				X
Sericea lespedeza				X	X
Annual lespedeza				X	
Alfalfa				X	
Birdsfoot trefoil				X	
Red clover		X	X	X	X
White/ladino clover		X		X	
Arrowleaf clover[2]	X				X
Berseem clover[2]	X	X	X		X
Crimson clover[2]	X				X
Hairy vetch[2]	X				X
Rose clover[2]	X				X
Subterranean clover[2]	X				X
Caley pea[2]		X	X		

[1]The X denotes compatibility under most conditions.
[2]Annual legumes such as arrowleaf, crimson, subterranean, and hairy vetch may be grown with tall fescue but are less desirable than perennial clovers.

A.30. Evaluating Alfalfa Stands

A. Plants Per Square Foot Needed for Optimum Yield

Age	Plants/square foot
New seeding	25–40
1st year	12–20
2nd year	8–12
3rd year and beyond	3–8

B. Stems Per Square Foot

>50	Excellent yields possible
30 - 50	Expect some yield reduction
<30	Significant hay and haylage yield reduction

A.31. Inoculation Groups for Commonly-Grown Southern Forage Legumes

Alfalfa Group
(Rhizobium meliloti)

Alfalfa
Black medic
Bur clover
Button clover
White sweetclover
Yellow sweetclover

Clover Group
(Rhizobium trifolii)

Alsike clover
Arrowleaf clover*
Ball clover
Berseem clover
Crimson clover
Hop clover
Persian clover
Red clover
Rose clover*
Subterranean clover*
White clover

Cowpea Group
(Bradyrhizobium japonicum spp.)

Alyceclover
Cowpea
Kudzu
Peanut

Lupine Group
(Rhizobium lupini)

Blue lupine
White lupine

Pea and Vetch Group
(Rhizobium leguminosarum)

Common vetch
Bigflower vetch
Hairy vetch
Roughpea
Winter pea

Other**

Birdsfoot trefoil *(Rhizobium loti)*

Soybean *(Rhizobium japonicum)*

*Specially selected strains specific for this legume species are most effective.
**Except for legumes listed under "Other" and those which require species-specific strains, the same inoculum can be used for all the species listed within an inoculation group.

A.32. Characteristics of Selected Plants for Wildlife

	Tolerance to soil acidity[1]	Annual or perennial	Fertility needs[2]			Tolerance to wet soils
			N	P	K	
Alfalfa	P	P	NA[2]	H	H	P
American jointvetch	F	A	NA[2]	M	M	P
Annual lespedeza	G	A	NA[2]	L	L	P
Browntop millet	G	A	H	M	M	F
Chufa	G	A	M	M	M	G
Egyptian wheat	P	A	H	M	M	P
Florida beggarweed	G	A	NA[2]	M	M	E
Foxtail millet	G	A	H	M	M	F
Japanese millet	G	A	H	M	M	E
Sorghums	P	A	H	M	M	P
Cowpea	G	A	NA[2]	M	M	P
Partridge pea	G	A	NA[2]	L	L	G
Proso millet	G	A	M	M	M	P
Sericea lespedeza	E	P	NA[2]	L	L	P
Sesame	P	A	L	M	M	P
Sesbania	G	A	NA[2]	L	L	E
Shrub lespedeza	E	P	NA[2]	M	M	P
Sunflower	F	A	L	H	H	P
Velvetbean	F	A	NA[2]	M	M	P
Annual ryegrass	F	A	H[3]	H	H	G

[1]E = Excellent; G = Good; F = Fair; P = Poor
[2]NA = Not applicable; properly inoculated legumes fix nitrogen which they use for growth. H = High; M = Medium; L = Low
[3]Winter annual grasses or grass/legume mixtures require applications of nitrogen for good growth; pure stands of legumes do not.

A.33. Animal Unit Equivalents

Animal Class	Animal Unit Equivalent[1]
Cow 1,000 lb non-lactating	1.0
Pregnant heifer ≥ 18 months	1.0
Bull > 24 months	1.5
Bull < 24 months	1.2
Cow and calf	1.3
Yearling > 18 months	0.9
Yearling 12-18 months	0.8
Calves < 12 months	0.6
Sheep and Goats	
Sheep and goat non-lactating	0.2
Ewe and goat with young	0.3
Weaned lamb or kid	0.15
Horse	
Draft	1.5
Saddle	1.25
Others	
Deer, fallow	0.17
Bison, mature	1.00
Elk, mature	0.65

[1]An Animal Unit is considered one mature non-lactating cow weighing 1,000 pounds and fed at maintenance level.

A.34. Common and Scientific Names and Life Cycles of Some Common Pasture Weeds in the Southeast

Common Name	Scientific Name	Life Cycle*
Annual foxtails	*Setaria* spp.	A
Bitter sneezeweed	*Helenium amarum*	A
Bracted plantain	*Plantago aristata*	A
Briars	*Rubus* spp.	P
Broadleaf signalgrass	*Brachiaria platyphylla*	A
Broomsedge	*Andropogon virginicus*	P
Buckhorn plantain	*Plantago lanceolata*	P
Cheat	*Bromus secalinus*	A
Common chickweed	*Stellaria media*	A
Common pokeweed	*Phytolacca americana*	P
Common ragweed	*Ambrosia artemisiifolia*	A
Crabgrass	*Digitaria* spp.	A
Curly dock	*Rumex crispus*	P
Cutleaf eveningprimrose	*Oenothera laciniata*	A
Dogfennel	*Eupatorium capillifolium*	P
Fall panicum	*Panicum dichotomiflorum*	A
Giant foxtail	*Setaria faberi*	P
Giant ragweed	*Ambrosia trifida*	P
Goldenrod	*Solidago* spp.	P
Hairy buttercup	*Ranunculus sardous*	A
Henbit	*Lamium amplexicaule*	A
Horsenettle	*Solanum carolinense*	P
Horseweed	*Conyza canadensis*	A
Italian ryegrass	*Lolium multiflorum*	A
Johnsongrass	*Sorghum halepense*	P
Little barley	*Hordeum pusilium*	A
Low hop clover	*Trifolium dubium*	A
Mouse-ear chickweed	*Cerastium vulgatum*	A
Musk thistle	*Carduus nutans*	B
Pennsylvania smartweed	*Polygonum pensylvanicum*	A
Pigweeds	*Amaranthus* spp.	A
Prickly sida	*Sida spinosa*	A
Red sorrel	*Rumex acetosella*	P
Rescuegrass	*Bromus catharticus*	A
Sandburs	*Cenchrus* spp.	A & P
Shepherdspurse	*Capsella bursa-pastoris*	A
Sicklepod	*Cassia obtusifolia*	A
Smutgrass	*Sporobolus indicus*	P
Swinecress	*Coronopus didymus*	A
Tall ironweed	*Veronia altissima*	P
Texas panicum	*Panicum texanum*	A
Torpedograss	*Panicum repens*	P
Vaseygrass	*Paspalum urvillei*	P
Virginia pepperweed	*Lepidium virginicum*	A
Wild garlic	*Allium vineale*	P
Wild mustard	*Sinapis arvensis*	A

*A = Annual, B = Biennial, P = Perennial

Index

Mention of a trade name within this book, or use of a photograph which includes a trade name or shows a commercial product, does not imply endorsement by the publisher, the authors, or the universities at which the authors are employed. Other products may be available which are equal or superior to any mentioned or depicted.
